Coal Desulfurization

Chemical and Physical Methods

Coal Desulfurization

Chemical and Physical Methods

Thomas D. Wheelock, EDITOR

Iowa State University

Based on a symposium

sponsored by the

Division of Fuel Chemistry

at the 173rd Meeting of the

American Chemical Society,

New Orleans, La.,

March 23, 1977

ACS SYMPOSIUM SERIES **64**

AMERICAN CHEMICAL SOCIETY

WASHINGTON, D. C.   1977

Library of Congress CIP Data

Coal desulfurization.
 (ACS symposium series; 64 ISSN 0097-6156)

 Includes bibliographic references and index.

 1. Coal—Desulphurization—Congresses.
 I. Wheelock, Thomas D., 1925-   II. American
Chemical Society. Division of Fuel Chemistry. III. Series:
American Chemical Society. ACS symposium series; 64.

TP325.C516   662'.623   77-17216
ISBN 0-8412-0400-4   ACSMC8   64 1-332 1977

ACS Symposium Series

Robert F. Gould, *Editor*

FOREWORD

The ACS SYMPOSIUM SERIES was founded in 1974 to provide a medium for publishing symposia quickly in book form. The format of the SERIES parallels that of the continuing ADVANCES IN CHEMISTRY SERIES except that in order to save time the papers are not typeset but are reproduced as they are submitted by the authors in camera-ready form. As a further means of saving time, the papers are not edited or reviewed except by the symposium chairman, who becomes editor of the book. Papers published in the ACS SYMPOSIUM SERIES are original contributions not published elsewhere in whole or major part and include reports of research as well as reviews since symposia may embrace both types of presentation.

CONTENTS

Preface ix

SULFUR IN COAL AND ITS DETERMINATION

1. Coal Microstructure and Pyrite Distribution 3
   Raymond T. Greer

2. The Determination of Forms of Sulfur in Coal and Related Materials 10
   John K. Kuhn

3. The Direct Determination of Organic Sulfur in Raw Coals 22
   B. Paris

PHYSICAL METHODS FOR REMOVING SULFUR FROM COAL

4. An Overview of Coal Preparation 35
   J. A. Cavallaro and A. W. Deurbrouck

5. Chemical Comminution: a Process for Liberating the Mineral
   Matter from Coal 58
   Philip H. Howard and Rabinder S. Datta

6. Use of the Flotation Process for Desulfurization of Coal 70
   F. F. Aplan

7. A Comparison of Coal Beneficiation Methods 83
   Seongwoo Min and T. D. Wheelock

8. Dry Table—Pyrite Removal from Coal 101
   D. C. Wilson

9. Magnetic Desulfurization of Some Illinois Basin Coals 112
   Haydn H. Murray

10. Desulfurization of Coals by High-Intensity High-Gradient Magnetic
    Separation: Conceptual Process Design and Cost Estimation 121
    C. J. Lin and Y. A. Liu

EXTRACTION OF SULFUR FROM COAL BY REACTION AND LEACHING

11. Applicability of the Meyers Process for Desulfurization of U.S. Coal
    (A Survey of 35 Coals) 143
    J. W. Hamersma, M. L. Kraft, and R. A. Meyers

12. Coal Desulfurization Test Plant Status—July 1977 153
    L. J. Van Nice, M. J. Santy, E. P. Koutsoukos, R. A. Orsini,
    and R. A. Meyers

13. Oxidative Desulfurization of Coal 164
    Sidney Friedman, Robert B. LaCount, and Robert P. Warzinski

14. Sulfur Removal from Coals: Ammonia/Oxygen System 173
    S. S. Sareen

15. Desulfurizing Coal with Alakline Solutions Containing Dissolved
    Oxygen 182
    C. Y. Tai, G. V. Graves, and T. D. Wheelock

16. Hydrothermal Coal Process 198
    Edgel P. Stambaugh

17. Coal Desulfurization by Low-Temperature Chlorinolysis 206
    George C. Hsu, John J. Kalvinskas, Partha S. Ganguli,
    and George R. Gavalas

    REMOVAL OF SULFUR BY PYROLYSIS, HYDRODESULFURIZATION,
    AND OTHER GAS-SOLID REACTIONS

18. Desulfurization and Sulfidation of Coal and Coal Char 221
    G. J. W. Kor

19. Fluid-Bed Carbonization/Desulfurization of Illinois Coal by the
    Clean Coke Process: PDU Studies 248
    N. S. Boodman, T. F. Johnson, and K. C. Krupinski

20. Hydrodesulfurization of Coals 267
    D. K. Fleming, Robert D. Smith, and Maria Rosario Y. Aquino

21. Improved Hydrodesulfurization of Coal Char by Acid Leach 280
    Ann B. Tipton

22. Coal Desulfurization during Gaseous Treatment 290
    Edmund Tao Kang Huang and Allen H. Pulsifer

23. Desulfurization of Coal in a Fluidized Bed Reactor 305
    G. B. Haldipur and T. D. Wheelock

Index 323

PREFACE

Among the various impurities in coal, sulfur is one of the most prevalent and pervasive, and it has long been a serious problem since it interferes in one way or another with most major uses of coal. In metallurgical coal, sulfur adds to the difficulty and cost of making steel because it must be kept out of the final product. In steam coal, sulfur compounds contribute significantly to the wear of pulverizing mills and conveying equipment and, furthermore, upon combustion produce sulfur oxides which corrode boilers and create an environmental problem. Since the rate of coal use for heat and power generation is projected to increase greatly during the next few years, the problems caused by the sulfur are also projected to increase.

Among various methods which have been proposed to solve these problems, precombustion desulfurization of steam coal offers several potential advantages over flue gas desulfurization, fluidized bed combustion, and conversion of coal into liquid and gaseous fuels. Precombustion desulfurization involves physical or chemical operations which span an entire range of complexity and sophistication. The simpler physical methods are the most cost effective, and even some of the proposed chemical desulfurization methods are likely to be more cost effective than processes which convert coal into liquid and gaseous fuels. Also precombustion desulfurization addresses not only the problem of environmental pollution but also the other problems of equipment wear and corrosion. In addition, desulfurized coal can be utilized in existing plants without extensive modification or retrofitting, and for non-base load power plants, the use of desulfurized coal can avoid tieing up investment capital in flue gas desulfurization systems which are not fully utilized. Furthermore in comparison with gaseous fuels, desulfurized coal is stored readily for future use.

In spite of these foreseen advantages and years of research and development, only a few desulfurization methods are highly perfected and used commercially. These methods in commercial use generally depend on making a simple physical separation of coarse sulfur-bearing mineral particles from coal. Unfortunately many coal deposits contain finely disseminated minerals and organically combined sulfur which cannot be removed by these methods. However, because of the impetus of government pollution control regulations, increasing demand for coal, and greater influx of both government and private sector research funds, the development of new and more effective methods for desulfurizing coal is accelerating. The new methods include a wide gamut of physical and chemical processes with many different workers and organizations involved in their development. As a result of this effort and an attractive cost–benefit ratio, it is anticipated that coal desulfurization will be increasingly important in the future, not only in preparing coal for direct combustion but also for coking and for conversion into liquid and gaseous fuels.

Because of the rising interest in coal desulfurization processes, a symposium on this subject was held which attracted a series of papers describing the results of recent research on a number of promising processes. These papers form the nucleus of the present volume. However, since not all areas of the field were equally well represented, additional material was added to produce a volume which provides a broad coverage of recent research. Included are reports of basic and applied research and development on physical separation methods (such as froth flotation, selective oil agglomeration, high-gradient magnetic separation, and dynamic segregation) and on chemical processes (such as leaching, chlorinolysis, pyrolysis, hydrodesulfurization, and gas-phase oxidation). There is also a review of both current and developing industrial coal preparation practices. Although these methods may alter the physical and/or chemical properties of coal and may produce liquid or gaseous by-products, the principal product is still a solid carbonaceous material. Hence, methods designed primarily to produce desulfurized liquid or gaesous fuels are not included.

Since progress in the field of desulfurization depends on knowing and understanding the nature and amount of sulfur in coal, several reports of work on characterizing the size distribution and physical form of the principal sulfur-bearing minerals (iron pyrite and marcasite) in coal and on developing new methods for determining the various chemical forms of sulfur in coal are also included. This related work is important because the total amount of sulfur in coal and its distribution between various physical and chemical forms vary greatly. Among the new developments discussed is a method for the direct determination of organically bound sulfur.

Because of its broad coverage and in-depth review of much of the current research on coal desulfurization, this volume provides both an introduction to the subject as well as a fairly comprehensive treatment and status report. Thus it should be of value to both the neophyte and the experienced worker not only because it describes possible solutions to the problem of sulfur in coal but also because it indicates where science and technology stand in achieving these solutions and where there is need for more research and development.

Many individuals contributed to the preparation of this volume. A number of researchers contributed the results of their efforts, the officers of the Fuel Chemistry Division of the American Chemical Society provided encouragement, and the editor and staff of the Symposium Series produced the volume. All of these individuals deserve our appreciation.

Iowa State University   THOMAS D. WHEELOCK
Ames, Iowa
October 3, 1977

Sulfur in Coal and Its Determination

Coal Microstructure and Pyrite Distribution

RAYMOND T. GREER

Department of Engineering Science and Mechanics, Engineering Research Institute,
Ames Laboratory, U.S. ERDA, Iowa State University, Ames, IA 50011

The microstructure of coal is of interest in understanding coal properties. Microstructural features may affect the selection of coal beneficiation methods for removing impurities. For example, in planning sulfur removal schemes it is useful to understand the proportionment of inorganic sulfur compared with that of organic sulfur. Details of size, shape, orientation, and distribution factors for pyrite, certain pyrite and maceral groupings or fields, or other coal constituents (macerals and other inorganic components) are of importance in supporting rational designs of sulfur removal processes. Recent scanning electron microscope (SEM) investigations help to show the interrelations between macerals and inorganic phases such as pyrite (FeS_2, the primary inorganic source of sulfur in coal) ([1,2,3,4]). This form of characterization complements optical microscope studies ([5,6,7,8]) and conventional transmission electron microscope (TEM) studies ([9-15]) which have been used to classify and to study coal constituents. SEM offers a way to visualize features hundreds and thousands of Angstroms in diameter by direct observation in three dimensions which is not available by any other technique. Many significant coal features (such as cellular components) and impurity crystals occur in this size range. Important features of coal constitution discovered by SEM are discussed below.

Experimental Investigation

Apparatus. A JSM-U3 SEM with an EDAX energy dispersive x-ray analysis system is used for the microscale structural and chemical studies. The resolution capabilities of the instrument on a day-to-day basis are somewhat better than 200 Å for microstructural information and submicrometer in a chemical analysis mode for localized crystal description. The system provides size, shape, orientation, and distribution information for inorganic phases and coal constituents.

Procedure. Samples for SEM studies are mounted in an orientation where the face and vertical position in the coal seam are
maintained. The electron beam accelerating potential usually
used in viewing the specimens is 25 kV. Samples receive a vacuum-
deposited coating of approximately 200 Å gold (for microstructural
studies; to minimize charging problems) or a vacuum-deposited
graphite coating of 20–100 Å (for microchemical studies; to avoid
possible interpretation difficulties of the sulfur and the gold
x-ray fluorescence lines). Additional information about experimental technique appears in Ref. 4.

Materials. Samples of high-volatile C bituminous coal from
both strip and shaft mines in Iowa form the primary materials of
interest in this work.

Results and Discussion

Microstructural Features. To see the interrelation of inorganic phases such as pyrite to the coal constituents (macerals),
micrographs have been chosen first to present the relative organi-
zation of the microstructural components. These will be followed
by additional micrographs representative of the common ways which
pyrite crystals occur in coal. Figure 1 provides details of
cellular compression. The coal specimen from the Lovilia (shaft)
mine is examined in two orientations: the top of a fresh fracture
surface corresponding to various magnification views of the top of
the coal seam (Figures 1a–c); and the side, end-on view of the
stratified material (Figures 1d–f). Fragmentation and compression
features of cellular material are seen for other coals in Figures
2, 3, and 4 and in Refs. 1 and 4.
    In Figures 1a–c, the coalified plant fibers run from left to
right in the images. Most of the fibers are approximately 25 µm
thick, as seen for example in a high magnification view in Figure
1c. Microstructural detail of pits and other individual cellular
fiber details can be discriminated in this view as well. The
openings and regions forming pathways of intercommunication for
fluids and gas between and within cells can be seen in side views
(Figures 1d–f) of the stacked fibers. As the resolution of the
scanning electron microscope is of the order of 150 to 200 Angstroms, micropores occurring below this size will not be observed,
even at the highest magnifications. Figure 1f shows the coal
openings in this case to be of the order of 20 µm or less, as influenced by the degree of compression of the fibers seen in the
particular field of view. Considerable compression of the plant
source material in the coal can occur, and in certain examples the
cellular detail may be completely absent down to the resolution
limit of this electron microscopy technique. The cell walls are
of the order of 2 µm thick in these side views (Figure 1f, for example).
    Figure 2 represents an example of both compression and dis-

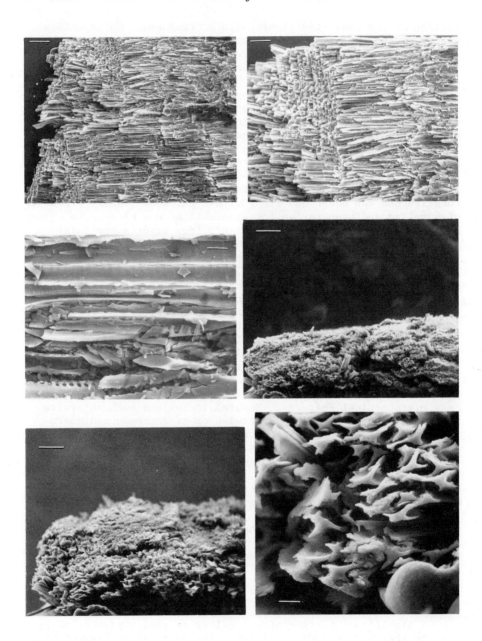

Figure 1.   Coal from the Lovilia mine (Iowa) showing detailed cellular features of the coalified plant material. (a) (top left) Top view; scale bar, 200 μm. (b) (top right) Top view; scale bar, 100 μm. (c) (middle left) Top view; scale bar, 20 μm. (d) (middle right) Side view; scale bar, 200 μm. (e) (bottom left) Side view; scale bar, 100 μm. (f) (bottom right) Side view; scale bar, 20 μm.

tortion of the coalified plant cellular features. The specimen
was tilted so that both the top and side of the specimen could be
seen in this view. Across the central region of the micrograph
this is particularly clear. An analogy would be to take a stack
of solid rectangular boxes (each open at the ends), each approximately 25 μm x 25 μm on the end by several hundred μm in length,
to represent a fiber, and crush and bend the stack by compression
perpendicular to the long dimension of the boxes. A result somewhat similar in appearance to Figure 2 could be obtained.

     Variations of cellular forms and fragments are seen in Fig-
ures 3 and 4. These represent fresh fracture surfaces of the coal
that reveal the material was already fragmented and crushed prior
to the scanning microscope investigation. This is a way the coal
constituents can occur and stack within the seam itself, exclusive
of sampling and preparation for examination on a microscale. Thus,
a wide range of detail may be present, from exact subcellular
features where little or no perturbation can be detected by microstructural examination, to that where there is a complete absence
of detail of the source material.

     These cellular openings can be filled with pyrite, gypsum,
calcite or other mineral inclusions which were deposited either at
the time of early stages of coalification or after the coal had
formed. In certain processing schemes to remove crystalline pyrite, the presence of this type of open micrometer size network
might be used to advantage depending in part on the occurrence,
distribution and surface characteristics of the type of coal constituent.

     <u>Microchemical Features</u>. Mineral inclusions such as pyrite
are revealed over a variety of scales of microscopic and macroscopic investigations. The sulfur-bearing phases are of particular interest in view of attempts to remove sulfur prior to combustion. The primary forms of sulfur in coal include:
     (1) Pyritic sulfur: FeS_2
     (2) Sulfate sulfur: $CaSO_4 \cdot 2H_2O$
     (3) Organic sulfur:
          a. mercaptan or thiol   RSH   (R and R' being alkyl
          b. sulfide or thio-ether   RSR'   or aryl groups)
          c. disulfide   RSSR'
          d. aromatic systems containing   HC — CH
              the thiophene ring   ‖   ‖
                                             HC   CH
                                               \ /
                                                S

     A significant portion of the sulfur in coal may occur as py-
rite in the form of individual crystals, or as assemblies of crystals forming framboids (a word coined by Rust (<u>16</u>) in 1935 meaning
berry-like), or as assemblies of framboids interconnected by additional pyrite occurring over a large-size range up to centimeters
in width or greater. Pyrite (FeS_2, cubic) and marcasite (FeS_2,

Figure 2. Compressed coalified plant cellular material showing both the top and the side of the fibers. Number 6, Midland, Ill. coal. Scale bar, 15 μm.

Figure 3. Plant debris. Mich mine (Iowa). Scale bar, 10 μm.

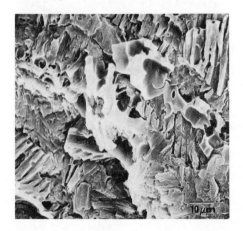

Figure 4. Plant debris. Both fragmentation and distortion are seen. Horton, KY. Scale bar, 10 μm.

orthorhombic) are observed in coal samples, the cubic form being
the most abundant of the sulfide minerals. Other inorganic forms
of sulfur such as gypsum are relatively low in abundance (<0.5 wt
%) or rapidly appear as weathering products from the pyrite (17).
There is little exact knowledge of the forms in which sulfur exists in organic combination. Organic sulfur is present in all
coal; it is determined as a numerical difference between total
sulfur and the sum of chemically determined sulfate sulfur (generally attributed to gypsum) and chemically determined pyritic
sulfur (18).

Representative forms of pyrite are shown in Figures 5a-k. In
these samples, micrometer-sized pyrite is an important constituent
of particles, nodules, concretions, and other pyrite assemblies.
The association of the pyrite with microstructural features of the
coal constituents can be observed in several of these micrographs.
For additional examples, see Refs. 1, 2, 3, and 4. Detailed pyrite size distribution data and distribution data for the forms
of pyrite with depth in a seam are presented in Refs. 1, 4, 19,
and 20. The single crystal form of pyrite predominates at approximately 1-μm particle diameters, but it is an important form over
a wide range of sizes. Commonly, micrometer diameter crystals
will be major constituents of large pyrite masses.

In Figures 5a and b, the occurrence of small individual pyrite crystals within sets of coalified plant fibers (telenite) is
presented at two different magnifications. The examination is
made in an orientation perpendicular to the top of the seam of
coal for this specimen from the seam (i.e., the top of a portion
of a seam of coal is seen). Most of the small bright specs seen
in Figure 5a are pyrite crystals. The central region of Figure 5a
can be examined at much higher magnification in Figure 5b. Here
the pocket of pyrite crystals in relation to the coalified plant
fiber is presented.

The pyrite crystals occur in other ways as well within the
coalified material. A horizontal opening in the coal (approximately 9 μm in height) forms the primary field of view in Figure
5c. The upper surface is covered with a layer of individual pyrite crystals. In Figure 5d, both individual separate pyrite
crystals and roughly spherical agglomerations ("framboids" ...
berry-like) of single crystals of pyrite fill an opening in the
coal (portions of the coal are seen in the upper left corner and
lower right corner of the photograph). In this example, notice
that there is space between the individual crystals and the framboid (the framboid is approximately 10 μm in diameter and is
located in the left center of the field of view; each of the crystals which comprise the framboid are approximately 1 μm in diameter). By comparison, Figures 5j and 5k represent a field of many
framboids where additional pyrite has filled in between the framboids and cemented them together to form a nodule or concretion
(the specimen from the mine was about a centimeter in diameter).
Another example of framboids having formed within an opening in

Figure 5. Forms of pyrite found in Iowa coal. (a) (top left) Deposit of micrometer-sized pyrite crystals within cellular features. Refer to Figure 5b also. Star mine. Scale bar, 20 μm. (b) (top right) Higher magnification view of central region of Figure 5a. Scale bar, 2.5 μm. (c) (bottom left) Pyrite crystals attached to upper surface of a horizontal opening in the coal. ICO mine. Scale bar, 1 μm. (d) (bottom right) Pyrite crystals occurring as a cavity filling. Both individual crystals and a spherical crystal assembly (a framboid) are seen. In this case, there is no infilling of material between the individual crystals and the framboid. Otley mine. Scale bar, 5 μm.

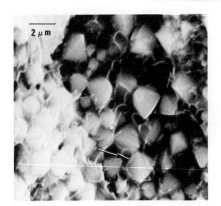

Figure 5. Forms of pyrite found in Iowa coal. (e) (top left) In the center of the field of view, a cluster of four spherical assemblies of pyrite crystals (four framboids) are seen. Star mine. Scale bar, 30 μm. (f) (top right) Infrequently observed polycrystalline sulfide. Otley mine. Scale bar, 3 μm. (g) (bottom left) Fresh fracture surface showing pyrite crystals within a coalified plant cell network. The arrow indicates a broken cell wall revealing a single pyrite crystal within this particular cellular feature. Compare this with Figure 5h. Mich mine. Scale bar, 2 μm. (h) (bottom right) Polished smooth surface of fusinite indicating coalified plant cell features (similar to a honeycomb appearance) with the roughly circular interiors filled with pyrite (or occasionally calcite, $CaCO_3$). Jude mine. Scale bar, 30 μm.

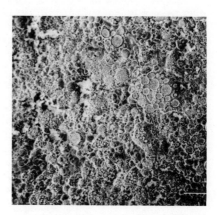

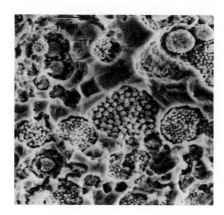

Figure 5. Forms of pyrite found in Iowa coal. (i) (top left) Gypsum crystals in fusinite. Jude mine. Scale bar, 4 μm. (j) (top right) Etched surface of a large (centimeter diameter) pyrite inclusion. Note spherical forms (framboids) and surrounding material (additional pyrite). Refer to Figure 5k also. Etching is necessary to reveal this. Mich mine. Scale bar, 50 μm. (k) (bottom) Higher magnification view of Figure 5j showing that each framboid is comprised of similarly sized individual crystals within a framboid. Mich mine. Scale bar, 10 μm.

coal is shown in Figure 5e. Figure 5f represents a pyrite inter-
growth feature.

     In certain cases (Figures 5g and 5h, for example), individual
crystals may be sufficiently large to completely fill coalified
plant fiber openings. To study this, flat polished sections are
of value. Using an electron microprobe x-ray analyzer with wave-
length dispersive analysis, two-dimensional surface distribution
analyses for elements such as carbon, sulfur, iron, and calcium
can be determined. The carbon displays can be associated with the
location of the coal, for example. The calcium displays (showing
the presence or absence of calcium) can help in locating calcite
($CaCO_3$) crystals. The combined occurrence of both iron and sulfur
is of value in identifying the presence and location of the pyrite crystals in relation to cell walls. Subsequently, scanning
electron microscope photographs of the regions examined by the
microprobe analyses can be obtained (Figure 5h, for example).

     Gypsum crystals of very small size (approximately 4 μm in
diameter by 8 to 20 μm in length) are shown in Figure 5i. Many
textural forms occur in coal; however, the amount of gypsum usually present is very low.

     Large masses of pyrite may be comprised of pyrite crystals
where the constituent pyrite crystal size may average less than a
micrometer in diameter. An example of this is presented in
Figures 5j and 5k. The entire field of view in these figures is
pyrite (the surface has been etched to reveal microstructural features). Here framboids comprised of individual crystals of the
order of a micrometer or less in diameter are present, and are
interconnected by additional pyrite. This is the result of a
multistage colloidal assembly process which formed the solid.
Within each framboid (Figure 5k, for example) the distribution of
sizes of the constituent individual pyrite crystals is small.
Average sizes of crystals among framboids are quite variable as
seen in Figure 5k. Figures 5j and 5k are cross-sectional views;
however, in Figure 5j, for example, there is a sufficient depth of
field so that the spherical forms of some of the framboids can be
seen even in such a cross-sectional examination (refer to the upper right central portion of the micrograph to see a few of these
clearly).

     Pyrite Distribution. Complete coal seam microscale chemical
and structural analyses have been performed for several channel
samples from Iowa strip and shaft mines (1,19). Using an energy
dispersive x-ray analysis system for point by point sulfur and
iron analyses together with a scanning electron microscope for
microstructural information (for studying crystal forms and coal
maceral-inorganic phase interrelationships) the distribution with
depth of pyrite crystals in channel samples from several mines has
been characterized. Also, for these mine samples the distribution
of pyrite among various forms of interest for planning pyrites removal schemes has been obtained as well as the proportionment of

TABLE I.   Distribution of Pyrite Among Various Forms

Form of Pyrite	Distribution, wt.%[a]		
	Mich mine	Star mine[b]	Lovilia mine[b]
Single crystal	69.7	21.1	7.5
Framboids	12.3	10.9	6.1
Other larger, massive forms	18.0	68.0	86.4
	100.0	100.0	100.0

[a]Fields of individual crystals are taken as size units; note, single crystals within these fields as well as isolated single crystals are included in the single crystal category.
[b]From Ref. 1.

pyrite in certain size ranges of practical interest. A summary of some of this information appears in Tables I and II.

In Table I, the data are presented in a convenient way where the pyrite is distributed among three forms in the seams (1,19): single crystals, framboids, and other larger, massive forms. The difficulty of separation of the pyrite from the coal by conventional washing and flotation methods is reflected in a high abundance of very small crystals and framboids, for example, such as seen in Figure 5.

Table II.   Proportion of Pyrite

Mich mine[a]
   50% < 12 μm
   75% < 76 μm (i.e., 200 mesh)

Mich mine[b]
   50% < 2 μm
   96% < 76 μm (i.e., 200 mesh)

Star mine[c]
   50% < 45 μm
   57% < 76 μm (i.e., 200 mesh)

Lovilia mine[c]
   50% < 300 μm
   30% < 76 μm (i.e., 200 mesh)

[a]With extensive fields of single crystals considered as size units.
[b]Single crystals considered on a constituent crystal basis.
[c]From Ref. 1.

An insight into the difficulty of pyrite removal by conventional separation techniques can be gained by examining proportionment data for the pyrite where the information is expressed as a weight (or volume) percent of pyrite less than a particular size (compared to the total pyrite determined for a channel sample). A brief summary is included in Table II. The 76 μm size is of potential interest since most of the coal burned for steam-raising is crushed to approximately 200 mesh.

Conclusions

The pyrite in coal occurs in several ways. The distribution of the forms of pyrite can be studied over an extended size range, to submicrometer sizes where necessary, on a basis of microstructure as well as element identification. This is of value in current investigations to assess the capabilities of coal desulfurization and recovery technology.

Acknowledgement

This work is supported at Iowa State University by the Engineering Research Institute and the Ames Laboratory under contract to the U.S. Energy Research and Development Administration under contract number W-7405-ENG-82.

Literature Cited

1. Greer, R. T., "Scanning Electron Microscopy/1977," I: 79-93, IIT Research Institute, Chicago, 1977.
2. Greer, R. T., "Recent Advances in Colloid and Interface Science," Vol. 5: p. 411-423, Academic, New York, 1976.
3. Greer, R. T., "Proc. Electron Microscopy Society of America," p. 620-621, Claitor's Press, Baton Rouge, 1976.
4. Greer, R. T., "Microstructural Characterization of Iowa Coal," IS-ICP-13, 1-76, June 30, 1975; IS-ICP-12, 1-14, May 31, 1975; IS-ICP-11, 1-20, Feb. 28, 1975; IS-ICP-10, 1-20, Oct. 3, 1974; available from Ames Laboratory, U.S. ERDA. Pub. Office, 201 Spedding Hall, Ames, IA 50011.
5. Augustyn, D., Iley, M., Marsh, H., Fuel (1976) 55: 25-38.
6. Boateng, D. A. D., Phillips, C. R., Fuel (1976) 55: 318-322.
7. Stach, E., "Stach's Textbook of Coal Petrology," 3-17, Gebruder Borntraeger, Berlin, 1975.
8. Spackman, W., Trans. N.Y. Acad. Sci., (1958) 20: 411-423.
9. Harris, L. A., Yust, C. S., Fuel (1976) 55: 233-236.
10. Leonard, J. W., Mitchell, D. R., "Coal Preparation," 1-44, 1-53, American Institute Mining, Metallurgical, and Pet. Engr., N.Y., 1968.
11. McCartney, J. T., O'Donnell, H. J., Ergun, S., "Coal Science," Adv. Chem. Ser. (1966) 55: 261-273.

12. Spackman, W., Barghoorn, E. S., "Coal Science," Adv. Chem. Ser. (1966) 55: 695-707.
13. Taylor, G. H., "Coal Science," Adv. Chem. Ser. (1966) 55: 274-283.
14. Brown, H. R., Taylor G. H., Nature (1962) 193: 1146-1148.
15. McCartney, J. T., Econ. Geol. (1949) 44: 617-620.
16. Rust, G. W., J. Geol. (1935) 43: 398-426.
17. Gluskoter, H., Energy Sources (1977) 3: 125-131.
18. Shimp, N. F., Kuhn, J. K., Helfinstine, J. K., Energy Sources (1977) 3: 93-109.
19. Greer, R. T., Proc. Int. Conf. X-ray Opt. Microanal. 8th (1977) in press.
20. Greer, R. T., "Proc. Electron Microscopy Society of America," Claitor's Press, Baton Rouge, 1977, p. 150-151.

2

The Determination of Forms of Sulfur in Coal and Related Materials

JOHN K. KUHN

Illinois State Geological Survey, Urbana, IL 61801

Standard methods of analysis for forms of sulfur in coal are adequate for geological mapping and present commercial usage. But because these procedures were intended for coal only, they will not give consistently correct values for coal-related materials. Additionally, as energy needs grow, the increasing use of coals of lower rank or of higher ash or sulfur content may present new unique problems in analysis. Direct determination of sulfides (other than as pyrite), organic sulfur, and elemental sulfur, is not a part of routine analysis for forms of sulfur in coal. Kinetic and mass balances for processing systems require not only these determinations but also a degree of accuracy heretofore unnecessary.

A recent review of the methods for determining sulfur and forms of sulfur in coal by Shimp et al. (1) considered thoroughly the existing procedures but did not discuss in detail the problems associated with unusual coals or coal-related materials. The following discussion of these problems as well as of current efforts to develop new methods may help the coal conversion industry to meet the demands likely to be placed on it in the future.

Standard ASTM Methods

Standard ASTM procedures (2) call for refluxing a coal sample for 30 min with dilute HCl, using a cold finger condenser, and then determining the sulfate sulfur by weighing the barium sulfate precipitated when barium chloride is added to the extract. The pyrite is extracted from the insoluble material remaining after the HCl leach with dilute HNO_3, and the pyritic iron is determined by titration with potassium dichromate or potassium permanganate. A recent revision in the method allows determination of the pyritic iron by atomic adsorption spectrometry, which may be the more accurate technique (3). Alternatively the HCl-soluble iron and the HNO_3-soluble iron may be determined on separate samples; the difference in the two values then

16

represents the pyritic iron. The organic sulfur is then determined by the difference between the inorganic sulfur (sulfate + pyritic sulfur) and the total sulfur in the original material, determined by the Eschka method (2), or by Eschka determination of the sulfur in the insoluble material from the HNO_3 extraction. If the organic sulfur obtained by difference exceeds the value determined on the residue, the former value is to be the accepted one. The major objection to this standard method is the fact that an error in any one step of the procedure can affect the values determined in other portions of the method.

Unusual Problems

The standard ASTM acid digestion procedure, for example, poses a potential problem. It has long been known that a number of coals contain finely dispersed pyrite, both crystallites and framboidal pyrite (4), that may be as much as 80 % of the total pyrite in the sample. This finely divided pyrite has caused little problem in analyses for different forms of sulfur because most of it can usually be dissolved easily, and errors of 0.1 - 0.2 % (absolute) are considerably less than the variations in sampling a coal seam or pile. However, some of the pyrite particles are small enough to be occluded in coal particles or to be encased in a kaolin lattice which can retard or prevent their dissolution in the standard ASTM acid digestion procedure, thus giving rise to small errors in the pyrite concentration obtained. Undissolved pyrite will cause errors in both the pyritic sulfur value and the organic sulfur value, which, even at the 0.1 % (absolute) level, will cause significant problems with kinetic or mass balance evaluation. The problem is not usually solved by grinding the sample more finely because these particles may be a few microns or less in size, and it is difficult to grind the samples sufficiently to insure total pyrite exposure.

Another source of error is the occurrence of sulfides (sphalerite, chalcopyrite, galena, etc.), other than pyrite, in many coals. While this occurrence is infrequent, investigations at the Illinois State Geological Survey have disclosed coals containing as much as 1.0 % sphalerite (ZnS) on a dry whole coal basis (5). When coals are processed or the mineral matter concentrated in some way, the concentration of these sulfides can exceed 10 % in the resulting samples. The acid digestion for sulfate sulfur in the ASTM procedure converts many of these sulfides to H_2S, which is evolved from the digestion flask and not retained for subsequent determination, therefore causing an error in the value of the organic sulfur content of the material. In addition, the concentration of various ionic species, which can be excessive in residues etc., may interfere in the titration for pyritic sulfur, but this problem is usually solved if atomic adsorption spectrometry is used to determine pyritic iron.

Another serious source of difficulty in the analysis of coal-
related materials is the occurrence of minerals not found normally
in coals in the residues and products of liquefaction and gasification processes. In the reducing systems often used to convert
coal to gas or liquid fuel, pyrite can be converted to pyrrhotite
(FeS). Sulfur is again lost as H_2S from pyrrhotite in a sample
dissolved in HCl for determination of sulfate sulfur, with a
resulting error in analysis. Other systems may change some of
the sulfur to the elemental form which is not separately accounted
for in the ASTM procedures. Extreme oxidizing conditions in some
systems, on the other hand, may yield large amounts of sulfates,
some of which are not very soluble in dilute HCl, and may give
rise to erroneous results.

Two other factors related to the analyses are important in
obtaining correct determinations of forms of sulfur. First, the
need for a representative and homogeneous sample is obvious, but
it can be very difficult to obtain because coal-related materials
can occur as liquids, solids, slurries, or highly viscous tars,
some with ash or mineral content ranging from near 0 to 100 %.
Often, extraneous materials such as filter aids, catalysts, etc.,
occur in the materials and are extremely difficult to remove or
to properly account for.

Second, the calculations necessary to compare the products
and residues from gasification and liquefaction processes to the
feed materials are affected by the mineral changes which have
taken place. For example, values are normally reported on an
ash-free basis, which works reasonably well for coals. Residues
from reducing liquefaction processes, however, can produce values
in excess of 100 % upon ashing. The reduced minerals can gain
weight in an oxidizing ashing system, and sulfur can be retained
in the ash in addition to being reported separately. Large
errors will be caused in the calculated values unless the results
are reported on the mineral-free basis (6).

New Methods

These problems have led a number of laboratories and
concerned groups, such as ERDA, EPA, and ASTM, to investigate
different approaches to determining forms of sulfur in coal and,
particularly, in related materials. The need for a standardized
method of analysis for comparing data between laboratories and
processes is evident. While such a method would not be necessary
for routine coal analysis, the present ASTM methods being
sufficient, it could serve as a special-purpose procedure.

Among the methods investigated is one adapted to coal (7)
from a technique for determining forms of sulfur in oil shale (8).
A strong reducing agent, lithium aluminum hydride, is added to
the coal in a closed system. After the pyrite is reduced to FeS,
the H_2S liberated on reaction with HCl is trapped in a cadmium
sulfate solution. This method can prevent the loss of sulfide

sulfur, since the sulfate sulfur can be removed after the pyritic
sulfur is determined. This procedure has several drawbacks, however. Organic sulfur is either calculated by difference of the
total sulfur and the inorganic sulfur values or determined on the
residue left after removal of the other forms of sulfur and is,
therefore, subject to the same errors as in the standard method.
The sample must be very finely ground to less than -400 mesh in
order to obtain the same level of precision as obtained with the
standard procedure. The method uses potentially hazardous
chemicals and is not suitable for use by an inexperienced analyst.
Finally, the pyritic sulfur value determined by this procedure
includes H_2S evolved from other sulfides. Despite these dis-
advantages, the method can, when used in conjunction with the
ASTM procedure, yield information about the amount of sulfides
other than pyrite present and about the concentration of
elemental sulfur if its level exceeds approximately 0.1 %.

Another approach to the problems is being used by the
Institute of Gas Technology (9). Their method closely follows
the ASTM scheme except that refluxing with HCl is done in a
closed system with the evolved H_2S collected in a solution of
cadmium sulfate and determined iodometrically. Sulfate sulfur is
determined as in the standard ASTM procedure. The pyritic iron
is extracted from the insoluble residue by refluxing with dilute
HNO_3, and sulfate sulfur is subsequently determined by atomic
absorption spectrometry. The nitric-acid soluble sulfur is
precipitated as barium sulfate; the difference between this value
and pyritic sulfur concentration is considered to be soluble
organic sulfur. This value for soluble sulfur is then added to
the value obtained for the sulfur in the residue by the standard
Eschka method. The sum of these two values represents the total
organic sulfur.

While these procedures do provide useful information, the
most interesting approaches attempt a more direct determination
of organic sulfur. Recent investigations at TRW (10), for
example, indicate that a low-temperature ashing technique can be
of significant value in determining forms of sulfur in coal-
related materials. The method assumes that a radio-frequency
generated plasma attacks selectively the organic constituents of
coal and related materials. Previous investigations by
Gluskoter et al. (11) have indicated that most minerals remain
unchanged in this process.

In the TRW procedure, coals or related products are low-
temperature plasma ashed to remove the organic sulfur as SO_x.
Total inorganic sulfur is then determined by x-ray fluorescence
analysis, the Eschka method, HNO_3 extraction, or another method
for the original material. This gives an organic sulfur value
reported to be, based on TRW's investigations, more precise than
that obtained by the standard method. Organic sulfur is still
determined by difference, but TRW is attempting to trap the SO_x
evolved and to determine it directly. A report on the precision

and accuracy of this approach is expected soon.

The use of plasma ashing for the direct determination of organic sulfur is also being investigated at Battelle's Columbus laboratories (12). This method is designed specifically for determining organic sulfur without interference from the other forms of sulfur present. The coal sample is plasma-ashed, and the resulting products (SO_2 + SO_3) are trapped in a cryogenic condenser and subsequently measured by ion chromatography. Results thus far show an increased repeatability over the results obtained by the ASTM procedure, and experiments with pyrite-"spiked" samples have shown no interference from large amounts of pyrite upon the determination of organic sulfur. This procedure works well with coal; however, the effect of minerals such as pyrrhotite is still being studied and may cause difficulties in the analysis of coal-related materials.

A method for the direct determination of elemental sulfur is being developed at Iowa State University (13). A l g sample of coal is refluxed with cyclohexane and the soluble sulfur determined by gas chromatography. Preliminary results reportedly indicate that 200 - 2000 ppm elemental sulfur may be present in various coals. While verification of these values is still necessary, they do not disagree significantly with information reported by Yurovskii (14).

Conclusion

Although these methods and others are still in the experimental stage, enough progress has been made that development of a standard method may soon be considered. This method should, if possible, include a direct determination of sulfide sulfur (other than pyritic sulfur), organic sulfur, and elemental sulfur. A working committee formed by EPA is considering the procedures with the intention of proposing and testing a method to serve the needs of developing process technology. These needs are also under consideration by ASTM; when a method of sufficient merit is presented, standardization of a procedure may be expected.

Literature Cited

1. Shimp, Neil F., Kuhn, John K, Helfinstine, Roy J., Energy Sources (1977) 3, (2), 93-109.
2. Book of A.S.T.M. Standards, (1976), Pt. 26, 269-283.
3. To be published in Book of A.S.T.M. Standards, (1978) Pt. 26, Revision approved by ASTM June 24, 1977.
4. Thiessen, R., Coal Age (1919) 16, (17), 668-673.
5. Ruch, Rodney R., Gluskoter, Harold J., Shimp, Neil F., Ill. State Geol. Surv., Environ. Geol. Note (1973), 61.
6. Givens, Peter H., Fuel (1976), 55, 256.

7.   Kuhn, John K., Kohlenberger, Larry, Shimp, Neil F., Ill.
     State Geol. Surv. Environ. Geol. Note (1973), 66.
8.   Lawlor, D. L., Fester, J. T., Robinson, W. E., Fuel (1963),
     42, (3), 239-244.
9.   Stotz, Robert, Institute of Gas Technology, personal
     communication (1977).
10. Kraft, Morton L., Hamersma, Warren W., T.R.W., personal
     communication (1977).
11. Gluskoter, Harold J., Fuel (1965), 44, 185-291.
12. Paris, Bernard, Battelle Columbus laboratories, personal
     communication (1977). (See also this volume.)
13. Richard, John, Vick, Ray, Junk, Gregor, Iowa State
     University, personal communication (1977).
14. Yurovskii, A. Z., "Sulfur in Coals", p. 80, translated from
     Russian 1974. Published for the U.S. Dept. of Interior,
     Bureau of Mines, and the National Science Foundation,
     Washington, D.C. by the Indian National Scientific
     Documentation Center, New Delhi. Available from: U.S.
     Dept. of Commerce, NTIS, Springfield, VA 22161.

3

Direct Determination of Organic Sulfur in Raw Coals

B. PARIS

Battelle, Columbus Laboratories, 505 King Ave., Columbus, OH 43201

The sulfur content among coals varies to some degree in total quantity and in the forms present. Generally, the forms of sulfur in coals are the inorganic (pyrite and sulfate) and the organic (a complex mixture of organo types). The sulfate content is usually present in low amounts except in some instances where weathering has converted some of the pyrite to sulfate. The pyrite form can occur in a rough size range from 0.25 to 200 μm.

The organic sulfur in coals is intimately bound to the coal molecule and is difficult to determine directly when other forms are present. The organic sulfur content in most American coals ranges from about 0.5 to 3%.

A total sulfur analysis will determine whether present SO_2 emission standards will be met upon combustion of the candidate coal. If the total sulfur is too high and stack gas cleaning is not available, some form of desulfurization is required before this energy source can be used. Cleaning processes can remove varying amounts of the inorganic forms, and some organic sulfur may also be extracted, but for all intents, the content of this latter form may be regarded as the lower level to which coal can be physically beneficiated. Therefore, the baseline level of organic sulfur must be known so that economic and time effective inorganic sulfur extractions can be applied which in turn can comply with combustion standards.

Current ASTM methods provide for the direct analyses of the total, sulfate, and pyritic sulfurs. The accepted techniques for determining the sulfur forms in coals are the ASTM (1) standard wet methods. The total sulfur is determined by the Eschka method (D3177-75) and the sulfate and pyrite by selective acid leaches (D2492-68). The organic sulfur is then obtained from the difference of the total and the two inorganic forms since no reliable direct method is available. The possible accumulation of errors and the time required (∿1 1/2 days) to obtain the organic sulfur by difference has made this approach undesirable but necessary.

The precision of the organic sulfur value obtainable from this approach was calculated from numerous analyses at Battelle-Columbus Laboratories (BCL) in conjunction with precision estimates of the respective ASTM methods. At two standard deviations (95% confidence level) the precision was found to be ±25%.

Instrumental methods also can be used to determine the inorganic forms and total sulfur for an estimation of the organic sulfur content. The two methods use X-ray analyses on pressed pellets of pulverized coal. In one, Hurley and White (2), used X-ray fluorescence to characterize all of the sulfur forms by wavelength shift of the SK_β peak caused by variation in the sulfur bonding. The second method (3) is discussed in an unpublished report on sulfur determination in coals and utilizes X-ray fluorescence for coal and sulfate sulfurs and X-ray diffraction for pyritic sulfide sulfur. Organic sulfur is obtained by the difference in the latter technique.

Both of the X-ray methods agree with corresponding wet chemical data on the same samples. The precision of these instrumental methods therefore is at least as good as the ASTM procedure, but the time of analyses is reduced to several hours. A direct method for determining organic sulfur in coals was reported by Sutherland (4) using the electron microprobe on pressed pellets of coal.

A BCL supported program, therefore, was initiated to develop a direct and specific method to determine the organic sulfur content in coals and in the presence of the other sulfur forms. The approach taken entailed the low-temperature reaction of ionized oxygen species with raw coals to yield volatile SO_x components which are the reaction products of the organic sulfur content of the coals. The SO_x products can be collected in a suitable trap maintained at low temperatures and subsequently analyzed.

Experimental Conditions

Equipment. For this purpose, the instrument used for the oxidation of coals during this program was the LFE LTA 600L low temperature asher. The generator operates at a crystal controlled frequency of 13.56 MHz and has an output capable of delivering 300 W of continuously variable power distributed among five separate reaction chambers. Normal ashing is carried out under continuous rough pumping (300 L/min capacity) so as to maintain a pressure of 1 torr at an oxygen input flow of about 50 cc/min. The low temperature oxidation technique has been used to study the mineral matter content of coals. As indicated previously, the technique utilizes the oxidative properties of the reactive ionic and atomic species of oxygen which are produced when molecular oxygen is passed through a high frequency field. The ionized oxygen species are directed to the surface

of the raw coal where selective oxidation of the organic content
leaves the mineral matter relatively unaltered for study. The
reaction produces volatile oxides of the total organic content.
These products are CO, CO_2, H_2O, SO_x, and NO_x, and they are
allowed normally to be discharged through the vacuum pump. The
temperature of the coal surface reaction under these conditions
can vary from 50° to 300° C depending primarily upon the power
applied to the RF coils. Frazer et al. (5) and Mitchell et al. (6)
used low temperature ashing of coals to study mineral stability
under a variety of conditions. Although better stability of
minerals was observed than that shown by the air oxidation
method at 400°C, it was noted that pure pyrite could be oxidized
at temperatures of 200 to 300°C.

To provide for the collection of the desired vapor species,
the exit of one of the reaction chambers was rerouted to accommodate a glass trapping system. This modification is shown
schematically in Figure 1. The all glass assembly, with ball
joint connectors includes a stopcock (D), a low temperature
trap (C) to collect the oxidized species, and a connector at (E)
to either evacuate during a run or to attach a gas bubbling trap
for collecting the SO_2 at the end of a run. The remaining
chambers were left intact so as not to interfere with the operation of the instrument.

Figure 2 shows the sample holder used for loading the sample
for the oxidation run. Sample plate (A) is placed in the holder
as shown after loading with coal. The sample holder containing
the sample is inserted in the modified sample chamber. This
overall arrangement provides ease of handling of the sample.

Reaction and Collection Procedure. A weighed sample is
loaded into the sample holder onto a glass plate. The sample
holder is inserted into the reaction chamber (A) through port
(B). The system is carefully evacuated to about 0.2 torr. As
the oxidation of the coals progress, the temperature (-196°C)
and pressure (1 torr) conditions permit trapping of the SO_2 and
SO_3 gases as solids.

The generator is turned off after, upon visual examination,
the samples appear to be completely oxidized. The oxidation
time can vary from 1 to 3 hours. Helium is slowly admitted
through stopcock (D) until the system can be opened to the air.
A glass impinger bottle containing about 50 cc of 3% H_2O_2
solution is attached to (E). With a slow purge of He through
the trap and impinger, the Dewar is removed to permit the transfer of the condensed SO_2 into the peroxide scrubber. The trap
is then rinsed with a peroxide solution to collect the SO_3
component of the product. The two solutions are combined for
sulfate analyses using the Dionex Model 10 ion chromatograph.

Samples. The coals which were selected for this program
were used as received except that they were ground to pass

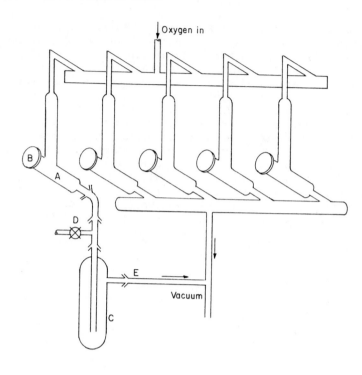

Figure 1. *Schematic of low-temperature asher showing modified reaction chamber and gas flows*

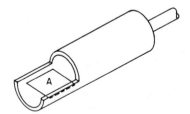

Figure 2. *Sample plate and holder insert*

-200 mesh sieve (74 μm). Wet chemical sulfur analyses were
obtained on these coals. In addition to a reproducibility study
on one coal, several other coals of varying sulfur and ash contents were run to establish the suitability of this method to
coals possessing varying chemical contents. The candidate
coals are listed in Table I along with their pertinent makeup.
The sulfur values were determined by ASTM methods.

The amount of coal samples taken for these runs was in the
20-30 mg range. The resulting sulfur collections were expected
to be of sufficient quantity to be determined by the analytical
method to be used. Small charges were used in an effort to
reduce the oxidation time to several hours. After the initial
surface of coal is removed, the penetration of the ionized
oxygen to the coal underlying the ash is time dependent. The
possible errors introduced in the overall determination of
organic sulfur from macroscopic inhomogenieties accentuated by
the small sample charges could be evaluated in the precision
series.

Table I.   Samples Used in Low Temperature Oxidation Studies

| | Sulfur (%) | | | | |
Sample	Total	Pyritic	Sulfate	Organic	Ash(%)
719-2 (Ohio)	6.3	1.58	2.74	1.9	14.2
719-3 (Ohio	5.2	3.6	0.07	1.4	17.1
Hazard #4 (Kentucky)	1.52	0.66	0.04	0.82	12.8
Colstrip #2 (Western)	0.68	0.18	0.14	0.36	9.0
Beach Bottom #1 (W.Va.)	1.97	1.35	0.03	0.59	25.1
Organic Compounds					
Dithiopropionic acid	17.7	0	0	17.7	0
2-Thiophene Carboxylic Acid	24.7	0	0	24.7	0

The pure organosulfur compounds shown were also run since
the results would represent ideal recovery experiments in the
absence of interfering mineral forms. The sample charges used
for these runs were in the 5-10 mg range.

A study was also made of the stability of FeS_2 under the
ashing conditions for coal. Oxidation of this compound yields
volatile SO_2 which would be indistinguishable from that derived
from the organic sulfur in coal. In addition, the pure FeS_2
was also mixed with other components to assess their effect
upon its conversion. These materials included pure SiO_2 graphite,
ash, and coal.

Experimental Results

Oxidation Study of FeS_2. It was first necessary to establish conditions which prevented or reduced considerably the oxidation of pyrite to yield SO_x. The experimental conditions which were varied for the FeS_2 study included applied power to the plasma, oxygen flow, position of samples in relation to the plasma, particle size of FeS_2, and mixtures of various materials with pure FeS_2. For the runs with only FeS_2, it appears that the conversion to Fe_2O_3 and volatile SO_2 occurs under any condition. However, the severity of conversion increases directly with the applied power and inversely with particle size. The position of the sample in the chamber and the rate of oxygen flow used have minimal, if any, effect upon the oxidation of FeS_2.

The runs made using mixtures of FeS_2 with the listed additives produced varied results. It was found that adding graphite, SiO_2, or an organic sulfur compound to pure FeS_2 did not prevent the FeS_2 reaction with oxygen. However, admixtures with coal or fly ash appear to prevent FeS_2 oxidation. The reason for the apparent inhibition of the reaction is not known but seems to be associated with characteristics of the ash other than SiO_2. The presence of the natural ash in coals accelerates the reaction of the organic phase with the plasma. In samples containing no ash, such as graphite or an organosulfur compound, the reaction with ionized oxygen is slower.

Coal Studies. Coals were oxidized using the optimum instrumental conditions as determined to be suitable by the pyrite studies. These were found to be (a) applied power of 150 W overall (20 W/chamber), (b) oxygen flow at 30 cc/min, and (c) a pressure of 1 torr. A sample charge of about 25 mg was used. The collected sulfates were determined on a Dionex Model 10 ion chromatograph.

Both chemical and system blanks were determined throughout the sample runs. The chemical blanks were obtained on equivalent volumes of peroxide-water solutions as used in the SO_x recoveries. The system blanks were determined from actual oxidation-transfer techniques without samples. Both blank levels were found to vary from 15 to 25 µg of sulfur.

Attempts were made to measure the temperature of coal surfaces under oxidizing conditions at 3 different RF power settings using a radiation thermometer. The surface temperatures corresponding to applied powers of 20 W, 30 W, and 40 W were 130°C, 210°C, and 230°C, respectively. There may be some doubt about the absolute accuracy of these values, but the relative differences should be satisfactory.

Using the above conditions for coal 719-2, a series of 10 runs supplied a measure of the repeatability of this technique for the organic sulfur recovery in one coal. These recovery

data and precision are shown in Table II. The ratio of the
obtained/expected values allows for the calculation of the relative standard deviation of 19% at the 95% confidence level. This
is slightly better than that achievable by the indirect ASTM
method.

The repeatability data shown in Table II include a wide
range of variables. These include: possible inhomogeneity of
small coal charges, ability to duplicate reaction and collection
techniques, analyses of sulfate solutions, and runs extended
over a period of several weeks.

Table II.   Repeatability Runs of the Analyses of Organic
            Sulfur From Oxidation of 719-2 Coal

Net Organic Sulfur (μg)		Ratio Obtained/Expected
Expected	Obtained	
440	450	1.023
405	410	1.012
420	475	1.131
420	360	0.861
380	360	0.949
430	450	1.047
440	440	1.00
415	380	0.916
430	350	0.814
Mean		0.971
% SD (95% confidence level)		19.0

It is noteworthy to point out the low recovery runs which
are represented in Table II by the ratio values 0.861 and 0.814.
Although the inhomogeniety of the sulfur distribution in the coal
could explain the overall variation of the results, another
possibility for the low recoveries could be the loss of SO_2
during transfer. More specifically, the loss could have occurred
during the helium purge into the evacuated system. An initial
high flow of helium could warm the deposited SO_2 sufficiently to
cause a slight vaporization loss to the system. At a temperature
of -196°C and a pressure of 1 torr, the calculated loss of SO_2
through vaporization could be 6 ng/hr. However, with the same
pressure conditions but at the higher temperature of -160°C, the
vaporization loss could be 30 mg/hr. Past experience has shown
that SO_2 cannot be condensed quantitatively at this pressure and
the higher temperature. Therefore, it is possible that for the
two low recovery values, the techniques of sample transfer were
inadequate.

The second study involving coal 719-2 was to oxidize for SO_x recovery after spiking with quantities of FeS_2 having a particle size range of <37 μm. The pyrite was premixed with coal at about the 12% level so that the charge to be oxidized was about 25 mg of coal and 3 mg of pure pyrite. The overall procedure for the mix runs was similar to that for the coal-only runs. Results of the coal-FeS_2 runs are shown in Table III. The relative standard deviation at the 95% confidence level is 17%.

Table III.   Analyses of Organic Sulfur from Oxidation of 719-2 Coal Spiked With Pure FeS_2

Organic Sulfur (μg)		Ratio Obtained/Expected
Expected	Obtained	
350	330	0.943
345	350	1.014
370	440	1.190
300	320	1.07
Mean		1.054
% SD (95% confidence level)		17.0

The third major experiment involved the repeatability of the described technique when applied to a wide range of coals. The basis for selection included a variation in total, pyritic, and organic sulfurs as well as the ash content. The coals used for these runs are listed in Table I.

The four coals were run in triplicate using similar procedure conditions as for the two previous studies with coal 719-2. The results for the four coals are shown in Table IV. One run (Colstrip) appears to be out of line, but the other ratio data seem to be well within the errors associated with either the ASTM or the present direct method. The results show that the direct (oxidation) method does provide a reasonable estimate of the organic sulfur content in a variety of coals.

Theoretical recoveries of sulfur in pure organic compounds were in the 90-95% range. The melting points of the two organosulfur components are near the temperatures achieved during the oxidations. It is uncertain what, if any, effect this has on full recoveries.

Observations. Other oxidized products of coals were condensed at the liquid nitrogen temperatures. Although no attempts were made to analyze the full contents of the condensed material, small amounts of ozone were present in runs which

contained organic carbon. There was no evidence of condensed
ozone when only FeS_2 or coal ash were exposed to the plasma.

Table IV. Application of the Oxidation Technique
to Various Coals

	Organic Sulfur (μg)		
	Expected	Obtained	Ratio
Colstrip #2	77	123	1.6
	75	85	1.1
	70	80	1.1
Hazard #4	220	173	0.79
	185	173	0.94
	200	190	0.95
Coshocton 719-3	330	300	0.91
	330	313	0.95
	325	340	1.05
Beach Bottom #1	180	195	1.1
	185	210	1.1
	170	190	1.1

Accidental leaks in the system during a run appear to
deplete not only the ozone, but also the condensed SO_2. The
resulting NO_x apparently reacts with SO_2 to form a NO-SO complex which makes it unavailable for the analytical method used.

Predeposition of SO_3 (probably as H_2SO_3 or H_2SO_4) on the
walls of the glass system can occur at high SO_x concentrations.
The deposition on the glass surfaces was very much in evidence
in the investigations with pure FeS_2. Blank levels were found
after coal runs.

Other SO_x collection techniques were tried in efforts to
simplify its recovery for analysis. One attempt included the
possible catalytic conversion of SO_2 to SO_3 for its higher
temperature collection. Poisoning of the platinum surface
occurred about halfway through the run to make this approach
unsuccessful. Another attempt involved placing a NaOH-laden
quartz wool plug and also ascarite in the rear of the holder
insert. Both appear to be very effective in capturing the SO_x
vapor. However, to be effective the NaOH quartz plug should be
slightly moist.

Conclusions

Based on the experimental work carried out for this program, a direct method for determining the organic sulfur content of coal has been developed. The procedure makes use of the relative ease of oxidation of the organically bound component of coals with ionized plasma of oxygen to yield volatile oxidized species peculiar to the total organic makeup. The desired products are condensed at a low temperature for subsequent analysis.

The work reported herein has demonstrated that this technique is specific for the organic sulfur content of coals within the error of the overall procedure. Repeatability runs on one coal (719-2) has produced relative standard deviations better than those obtained for organic sulfur using the different methods of ASTM. Similar precision was obtained from the same coal even when spiked with comparatively large quantities of pure FeS_2 having a particle size of <37 µm. No sacrifice in the amount of organic sulfur recovered was indicated in these spiked runs.

It has also been shown that this technique is suitable for coals of varying makeup. In addition, recent success in determining organic sulfur in solvent and chemically refined coal products provides a wider base of applicability of the method.

The repeatability runs indicate that the use of the 25 mg sample charge is justified. Any error from the inhomogeneous distribution of organic sulfur in the use of small charges appears to be minimal or within the expected limits of precision of the method.

Literature Cited

1. "Book of ASTM Standard Methods Part 26", D3177-75, D2492-68 (1976).
2. Hurley, R. G., White, E. W., "New Soft X-ray Method for Determining Chemical Forms of Sulfur in Coal", Anal. Chem. (1974), 46, 2243.
3. Paris, B., Schumacher, P. W., "An XRF-XRD Method for Determining the Total and Inorganic Sulfur Quantities in Coal", Proceedings of the Symposium on Electron Microscopy and X-ray Applications to Environmental and Occupational Health Analyses (April 25-27, 1976), Denver, Colo. (To be published).
4. Sutherland, D. C., "Determination of Organic Sulfur in Coal by Microprobe", Fuel (1975), 54, 41.
5. Frazer, F. W., Belcher, C. B., "Quantitative Determination of the Mineral--Matter Content of Coal by a Radio Frequency Oxidation Technique", Fuel (1973), 52, 41.
6. Mitchell, R. S., Gluskoter, H. J., "Mineralogy of Ash of Some American Coals: Variation with Temperature and Source", Fuel (1976), 55, 90.

Physical Methods for Removing Sulfur from Coal

An Overview of Coal Preparation

J. A. CAVALLARO and A. W. DEURBROUCK

Coal Preparation and Analysis Laboratory, Bureau of Mines,
U.S. Department of the Interior, Pittsburgh, PA 15213

Current Coal Preparation Practices (1)

In the United States today coal preparation practices are
directed toward maximizing Btu recovery, 3-shift-per-day operation of preparation plants, fine-size coal washing, closed water
circuits, and reducing sulfur in the final product.

In 1976, about 665 million tons of coal were produced, an
increase of 24 million tons from 1975. Table I compares clean
coal tonnage produced by mechanical cleaning and thermal drying
equipment in 1975. Dense-medium processes produced 32.6% of all
the clean coal, an increase of 1.6%, while the percentage of
coal cleaned by jigs dropped 2.2% to 46.6%. Although only 2.5%
of the coal was produced by dry cleaning, interest is growing in
the application of dry separation techniques to remove impurities
from western coal deposits where water is limited.

The tonnage of thermally dried coal continues to decrease as
increasingly stringent particulate emission regulations escalate
the cost of thermal drying. Since 1970 the proportion of thermally dried bituminous coal has decreased 50%. However, the use
of thermal drying will increase in the future as more of the low-
sulfur western coals with high inherent moisture of about 25% are
used to meet the increased demand for low-sulfur coal. Fluidized-
bed dryers now dry 72.5% of all the thermally dried coal.

In the United States the average coal preparation plant is
operated only about 13 hr. per day, 5 days per week. A one- and
two-shift-per-day operation is being questioned as the cost of
coal preparation increases owing to inflation and as the new plant
circuits being developed become more complex to meet exacting
ash and sulfur levels with minimum loss of Btu's. Consequently,
a number of the new plants are being designed to wash coal 24 hr.
per day. Where equipment must be out of service for periodic
maintenance, such as pumps that move abrasive material, backup
units are being installed. These new plants normally will have
capacities of 1000-1500 tons per hr. and will have two or more
parallel coal washing circuits. In this way, one circuit can be

Table I. Mechanical Cleaning and Thermal Drying of
          Bituminous Coal and Lignite, by Type of
          Equipment, 1975

Type of equipment	(thousand short tons)
Cleaners	
Wet methods	
Jigs	124,317
Concentrating tables	28,682
Classifiers	6,176
Launders	2,664
Dense-medium processes	
Magnetite	72,448
Sand	13,533
Calcium chloride	951
Total[a]	86,932
Flotation	11,518
Total, wet methods[a]	260,289
Pneumatic methods	6,704
Grand total[a]	266,993
Dryers	
Fluidized-bed	25,866
Multilouver	1,969
Rotary	794
Screen	2,798
Suspension or flash	4,184
Vertical tray and cascade	70
Total[a]	35,681

[a]Data may not add to totals shown because of independent rounding.

down for maintenance while the other is processing coal, thus
avoiding a complete shutdown of the coal washing facility.   In
this way, plant operating costs can be considerably reduced.

Sulfur and Its Removal.   Sulfur in coal occurs in three
forms--organic, sulfate, and pyritic.   Organic sulfur, which is
an integral part of the coal matrix and which generally cannot be
removed by direct physical separation, comprises 30-70% of the
total sulfur of most coals.   Generally, the organic sulfur-total
sulfur ratio is highest for low-sulfur coals and decreases as the
total sulfur content increases.

The sulfate is normally an oxidation product which is water
soluble and therefore readily removed during coal cleaning.   Sulfate sulfur content is usually less than 0.05%.

Pyritic sulfur in the form of mineral pyrite occurs in coal
as discrete and sometimes microscopic particles.   It is a heavy
mineral with a specific gravity of about 5.0 whereas coal has a
maximum specific gravity of only 1.8.   The pyritic sulfur content
of a coal generally can be reduced significantly by size reduction
and subsequent specific gravity separation.

Figure 1 shows the results of stage crushing from 1 1/2-inch
top size to 200 mesh top size and gravimetrically separating a
sample of Lower Freeport bed coal from Ohio and a sample of Pittsburgh bed coal from West Virginia.   For the Lower Freeport bed
coal sample, crushing from 3/8 inch to 14 mesh provided the maximum change in pyritic sulfur reduction for the float-1.30-specific-
gravity material and also for the float-1.60-specific-gravity
material.   For the Pittsburgh bed coal sample, maximum change in
sulfur reduction of about 0.8 percentage point was obtained at
either specific gravity by crushing from 14 to 48 mesh.   Interestingly, further crushing of the Pittsburgh bed coal sample from 48 to
200 mesh showed an additional pyritic sulfur reduction of about
0.5 percentage point for the float-1.60-specific-gravity material.
Regardless of how difficult and expensive the processing of coal
crushed to 48 or 200 mesh top size is, the potential benefits of
obtaining low-sulfur products may well justify the additional
costs and problems encountered.

Controlling Sulfur Oxide Emissions.   Available methods for
controlling sulfur oxide emissions from stationary combustion
sources fall into the following major categories:
    (1)   The physical removal of pyritic sulfur by coal
cleaning prior to combustion.
    (2)   The removal of sulfur oxides from the combustion
flue gas.
    (3)   Conversion of coal to a clean fuel by such
processes as gasification, liquefaction, and
chemical extraction.
Of these methods, physical removal of pyritic sulfur is the
least expensive and most highly developed.   The degree of sulfur

reduction possible depends on the washability characteristics of
the raw coal and its amenability to sulfur release upon crushing;
these characteristics are unique and vary from coal to coal.

Sulfur Reduction Potential Studies. In 1965, NAPCA (EPA)
funded a study to determine washability characteristics of U.S.
coals. The purpose of this continuing study is to determine the
forms of sulfur in the major sources of utility steam coals and
the sulfur reduction potential of these coals. The information
generated by this study is necessary to assess the impact physical
desulfurization of coal might have on the level of sulfur oxide
emissions from stationary combustion sources.

A Bureau of Mines report entitled "Sulfur Reduction Potential
of the Coals of the United States" (2) summarizes the work performed from 1965 to mid-1974. This report presents the results
of a washability study of 455 raw coal channel samples with special emphasis on sulfur reduction.

The 455 U.S. coal samples evaluated contained on the average
1.91% pyritic sulfur and 3.02% total sulfur.

Of the 455 samples tested, 174 either required no beneficiation or could be upgraded to meet the EPA new source emission
standard of 1.2 lb SO_2/MM Btu. The Btu recovery of these samples
averaged 88.7% and ranged from 5.6 to 100%.

If only those coals with a minimum Btu recovery of 50% are
considered, then 36% of the 455 samples tested would meet the
new standard with an average Btu recovery of 91%.

Effects of Beneficiation on Coal Reserves (3). The coal
resources in the States east of the Mississippi River containing
1.0% sulfur or less are estimated at 95.3 billion tons (Figure 2).
Although this is only 20% of the total remaining resource, it is
still a significant amount. Available washability data indicate
that much of the medium-sulfur-content coal can be upgraded to
1.0% sulfur or less using state-of-the-art coal cleaning
techniques.

An effort is now underway to determine how the present reserves of coals east of the Mississippi River would be categorized
in terms of sulfur content if they were subjected to different
levels of coal beneficiation. This involves an overlay of the
sulfur reduction potential data and the reserve base data generated by the Bureau of Mines Eastern Field Operations Center. This
study has been completed for the Northern Appalachian Region, where
a dramatic increase occurs in the projected reserves of low-sulfur
coal.

The tonnages, by sulfur category and extent of preparation,
for the entire Northern Appalachian Region are given in Table II.
The present reserves with sulfur content less than 0.85% amount
to 1,727.45 million tons. Crushing to 1 1/2 in. top size and
washing at 1.60 specific gravity yields 4,172.32 million tons of
low-sulfur coal, which is 2.4 times the existing reserves.

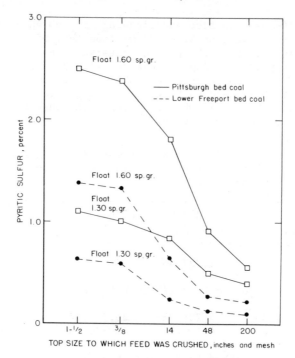

Figure 1.   Pyritic sulfur reduction potential for two Appalachian region coals as a result of stage crushing

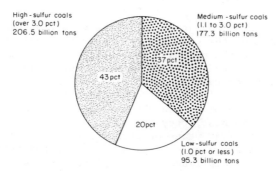

Figure 2.   Estimated remaining coal resources of all ranks, by sulfur content, in states east of the Mississippi River, Jan. 1, 1965

Crushing to 14-mesh top size and cleaning at 1.60 specific gravity
would result in a total of 8,546.31 million tons of low-sulfur
coal, which is 4.9 times the present reserves. Four percent of
the Northern Appalachian Region reserves with a sulfur content
greater than 0.85% can be upgraded to low-sulfur coal by crushing
to 1 1/2-in. top size and washing at 1.60 specific gravity; the
percentage increases to 10 when crushing to 14-mesh top size and
washing at 1.60 specific gravity.

Table II. Coal Preparation Wins Low-Sulfur Coal for Energy
           Market in Northern Appalachian Region, Millions
           of Tons

Product	<0.85% Sulfur	0.85-3.00% Sulfur	>3.0% Sulfur
Raw coal	1,727.45	38,995.41	27,535.04
Crushed to 1 1/2-inch top size			
Cleaned at 1.30	4,237.93	20,398.17	637.26
Cleaned at 1.60	4,172.32	43,396.58	14,164.45
Crushed to 3/8-inch top size			
Cleaned at 1.30	6,485.79	22,350.99	651.68
Cleaned at 1.60	6,224.53	44,595.82	9,918.26
Crushed to 14-mesh top size			
Cleaned at 1.30	7,136.62	21,691.77	395.59
Cleaned at 1.60	8,546.31	45,873.05	5,956.40

Figures 3 and 4 show graphically the effect of crushing to
various top sizes on the tonnages of low-sulfur coal potentially
available in the Northern Appalachian Region. In examining these
figures to establish the benefits of coal preparation, it is
necessary to keep several things in mind. First of all, even
though the tonnages of low-sulfur coal (less than 0.85% sulfur)
resulting from coal preparation are a relatively small fraction
of the total reserve base, they represent sizable quantities of
coal, equivalent to many years of production at current and foreseen rates. Moreover, many powerplants constructed prior to 1975
can be brought into compliance with local emission standards by
using coal cleaned to the appropriate sulfur level.
     In summary, coal preparation has the potential of providing
sizable tonnages of coal that will meet the EPA new-source

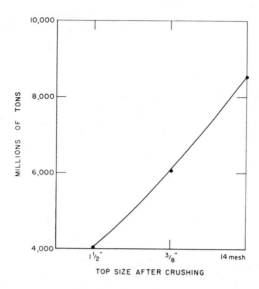

Figure 3.   Effect of crusing and cleaning at 1.60 specific gravity on the availability of < 0.85% sulfur coal

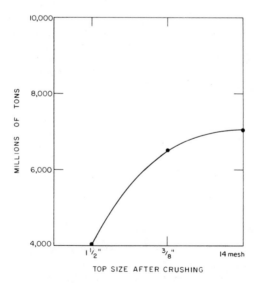

Figure 4.   Effect of crushing and cleaning at 1.30 specific gravity on the availability of < 0.85% sulfur coal

emission standards. In addition, for powerplants built prior to
1975, far greater quantities of coal can be burned after upgrading
by coal preparation.

New Developments in Coal Preparation (4)

Coal preparation state-of-the-art is changing rapidly in the
United States today. The scope of these changes and some of the
new circuit and equipment innovations which have been recently
introduced are discussed below.

The major emphasis is on reducing sulfur and maximizing Btu
recovery. The need for sulfur reduction stems from the enforcement of SO_2 emission standards, while high Btu recovery becomes
essential as a result of the increased value of coal as an energy
source. The resulting irony spells out the challenge facing coal
preparation today: How to economically produce increasing amounts
of coal that satisfy the strict pollution standards.

Homer City Preparation Plant. A possible solution to the
dual problems of environmental restrictions and maximum recovery
is the highly publicized Homer City multistream coal cleaning
plant (Figure 5). The plant is being built for joint owners,
New York State Electric and Gas Corp. and Pennsylvania Electric
Co., by Heyl & Patterson, Inc. The plant will produce two
streams of clean coal. The first stream will be an extremely
high quality product which will meet the environmental standards
for firing in the new Unit 3 generating station (650 MW) without
subsequent stack gas scrubbing.

The second clean coal stream is of intermediate quality and
will meet the existing source requirements for Units 1 and 2,
which are 600 MW each. The expected preparation plant performance, given in Table III, shows that the sulfur content is reduced
about 70% from 2.8% to 0.88% in the Unit 3 clean coal while the
overall Btu recovery is 94.5%.

Table III. Homer City Preparation Plant Expected Performance[a]

	Units 1 & 2	Unit 3	Refuse
Recovery (weight percent)	56.2	24.7	19.1
Recovery (Btu percent)[b]	61.6	32.9	5.5 (Loss)
Product Btu/pound (dry basis)	[c]12,549	15,200	3,367
Ash, percent	17.8	2.8	69.7
Sulfur, percent	2.84	0.88	
Pounds sulfur/MM Btu	1.79	0.58	

[a]Reference: J. F. McConnell, Homer City Coal Cleaning Demonstra-
    tion, 4th National Conference on Energy and Environment (EPA).
[b]Gross recovery, 94.5%; net recovery is 93.5% with losses for
    thermal dryer included.
[c]Blend of middling and clean coal.

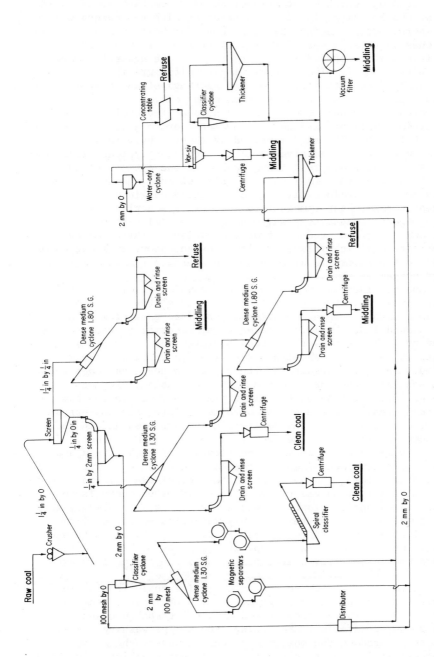

Figure 5.   Flow diagram of Homer City preparation plant

The plant includes several innovative applications of existing technology which we at the Bureau of Mines have been researching at a pilot plant level in cooperation with the owners and designers. In producing the clean coal for Unit 3, dense-medium cyclones clean at a specific gravity of separation of 1.3; a specific gravity of 1.8 is used to reclean the underflow material for Units 1 and 2. These separating specific gravities are significantly beyond the normal operating range of about 1.4-1.6 for dense-medium cyclones washing bituminous coal.

Also, the usual 28-mesh bottom size of the coal cleaned in the cyclone will be extended to 100 mesh, or to about one-fourth the usual size. The feed is sized at 1/4-in. by 9-mesh and 9- by 100-mesh fractions and cleaned in two separate cyclone circuits, instead of the whole feed being washed in one cyclone circuit.

Because of the small particle size of coal cleaned in the 9- by 100-mesh cyclone circuit, conventional drain and rinse screens for subsequent magnetite recovery are replaced by two-stage magnetic separators which handle the full streams of overflow and underflow products from the cyclone. The cleaning of finer size coal in cyclones followed by the above application of magnetic separators is also being considered for installations by other companies.

Another interesting feature is a novel control circuit for close specific gravity control of the feed slurry to the dense-medium cyclones. When operating at specific gravities of separation around 1.3, the ±0.10 near-gravity material in the feed is about 60%. Therefore, the entering slurry must be rigidly maintained at the proper specific gravity or material will be misplaced. This will result in either a clean coal that does not meet specifications or a decrease in high-quality clean coal tonnage. Figure 6 shows the pilot plant control system that is now being tested at the Bureau of Mines. Three streams—fresh water, coal slurry, and magnetite slurry—are combined in the mixing sump and represent the feed to a dense-medium cyclone. Instruments measure the density and flowrate of each stream and send the measurement signals to an analog computer. The computer calculates these signals, compares the actual density and flow for the feed slurry with the preset values, and then controls the water and magnetite slurry flowrates accordingly to produce the preset values. The system not only maintains strict quality control at a high efficiency, but provides the flexibility needed to make quick changes in the specific gravity of separation, should the washability characteristics of the raw coal change.

Batac Jig. The trend toward fine-size coal cleaning to remove more sulfur and recover more Btu's has spurred the development of closed water circuits and many new techniques and circuits in washing, dewatering, and sizing. Baum jigs washing coarse coal still produce the most clean coal; however, the Batac jig, developed in Germany to clean fine coal, has been installed in at

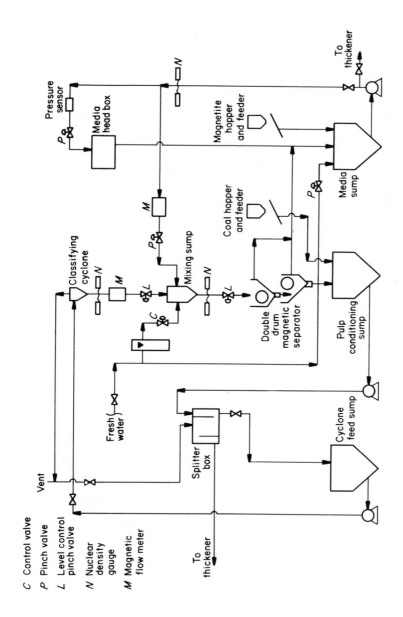

Figure 6.   Dense-medium specific gravity control circuit

least three U.S. plants to process 1/2-in. by 0 raw coal. It is
generally installed after a Baum jig which cleans the plus 1/2-
inch raw coal. In the Batac jig, air pulsations are produced in
multiple chambers directly beneath the bed screen, as shown in
Figure 7. The pulses in each chamber are controlled independently
and precisely. The jig provides significantly higher throughput
in the same space as a conventional Baum jig because of its
improved efficiency and elimination of the adjacent air chamber.
The Batac jig cleans with a sharpness similar to that of concentrating tables for the same size range.

Novel Fine-Size Cleaning Circuit. An interesting new fine-
size coal cleaning circuit has been installed in a Pennsylvania
preparation plant by Roberts & Schaefer Co. The plant was originally strictly an air table plant, but a dense-medium cyclone has
been incorporated to reclean the second and third refuse draws
from the air table. The first draw is essentially pure refuse
with no misplaced float material. With the second and third draws
wide open, the amount of sink material reporting to the clean coal
is reduced. At these table settings, about 20% of the feed is
sent to the cyclone to maximize Btu recovery while upgrading the
final product. The cyclone produces a 5.6%-ash clean coal, which
is combined with the clean coal from the air table. The net results are a 5% increase in yield and a 20% decrease in ash content. The air table is receiving attention from western coal
producers who are looking for a dry method of cleaning coal in
the water-scarce Western States.

Dewatering and Sizing. One of the biggest problems in
processing fine-size coal is dewatering and sizing. The severe
blinding tendency of minus 28-mesh coal slurries on conventional
screens makes for extremely inefficient and unreliable separations. A new screening device has been developed by Derrick
Manufacturing Corp. which purports to have solved the blinding
problem. It consists of a "sandwich screen" concept, which binds
two fine screen surfaces one on top of the other. The action of
the two screens against each other, from a special high frequency-
low amplitude vibrating motor, prevents near-size particles from
becoming lodged in the top screen. The openings in the bottom
screen are slightly larger than those in the top to ensure that
particles are not trapped between screens. Derrick claims these
screens can make sharp separations down to 400 mesh. The final
section of the screen may be a slotted natural rubber surface
fitted with a patented vacuum arrangement to help remove additional surface moisture.

Closed Water Circuits. In addition to the enforcement of
air emission standards, stream pollution and refuse disposal laws
have become more stringent recently. Thus, new preparation plants
use some type of closed water circuit with a gravitational

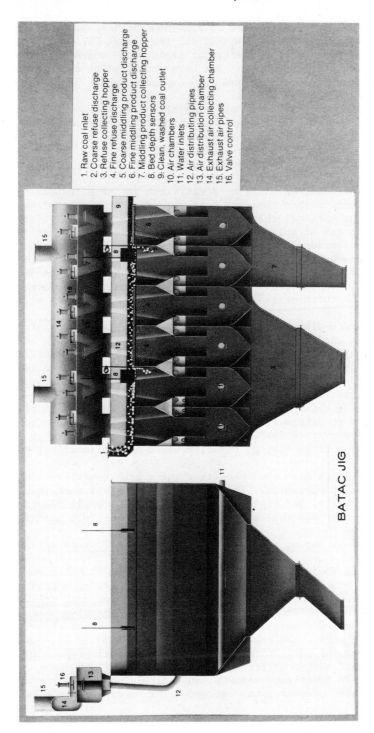

1. Raw coal inlet
2. Coarse refuse discharge
3. Refuse collecting hopper
4. Fine refuse discharge
5. Coarse middling product discharge
6. Fine middling product discharge
7. Middling product collecting hopper
8. Bed depth sensors
9. Clean, washed coal outlet
10. Air chambers
11. Water inlets
12. Air distributing pipes
13. Air distribution chamber
14. Exhaust air collecting chamber
15. Exhaust air pipes
16. Valve control

BATAC JIG

Figure 7.   Batac jig

clarifier as the primary piece of equipment in the system. Two
new thickener designs—the Lamella and the Enviro-Clear—are of
interest for possible application at plant sites where land area
is at a premium. These units have been installed at older plants
which had space limitations and were expanding their plant capacity or eliminating waste ponds.

Both compact thickeners may require as little as one-tenth of
the space occupied by a conventional circular tank thickener. The
clarity of their overflow is competitive with that of conventional
thickeners with overflow concentrations of 500 to 100 ppm or less
of solids, while underflow concentrations are slightly higher at
30-40% solids.

Each thickener operates on a different design principle. The
Lamella (Figure 8) has the same settling area for a given flowrate as a conventional thickener by virtue of a series of plates.
The settling out of solids on these plates is much like that in a
stack of conventional thickeners, but the plates are inclined to
allow continuous removal of the solids. Lamellas have feed loading rates of 0.5 to 0.6 gpm/ft^2, and use flocculant dosages of
about 2 ppm. These values are in line with those of the large
circular tank thickeners. At present, the largest Lamella will
handle 1,500 gpm.

The Enviro-Clear (Figure 9) has a much higher feed loading
rate, 3 to 4 gpm/ft^2, because it does not have the settling area
of the Lamella thickener. It is a bottom-fed unit which actually
uses the settled solids as a filtering medium for the water as it
flows upward to the overflow. There are only two zones in this
thickener, clarified liquid and compressed solids. Feed fluc-
uations can be absorbed by allowing the sludge bed level to rise
or fall with no detriment to the overflow clarity. Both units
are more sensitive to feed variations, and they lack the huge
advantage of a large storage capacity which conventional thickeners have.

The examples discussed show that there are innovations in all
areas of coal preparation. The new equipment, novel techniques,
complex circuitry, and higher levels of control are all needed in
order to insure that coal market demands can be met economically.

Summary of U.S. Bureau of Mines Coal Preparation Program (5)

Prior to 1965, the main thrust of the Bureau of Mines coal
preparation research effort was directed toward ash reduction with
little effort aimed at the environmental aspects of coal utilization. The research impetus clearly mandated by the Clean Air Act
of 1970 is that SO_x emissions must be controlled, trace elements
present in the raw coal must be identified and tracked to final
disposal, and the environmental effects of the coal preparation
process itself must meet stringent requirements.

With this legislated mandate, an entirely new dimension of
coal preparation research has emerged. The Bureau of Mines has

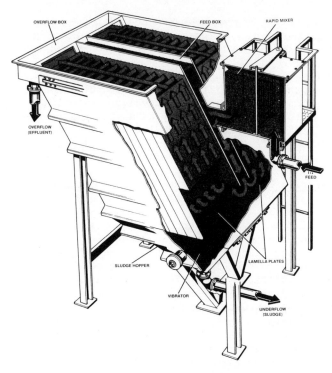

Figure 8.   Lamella
thickener

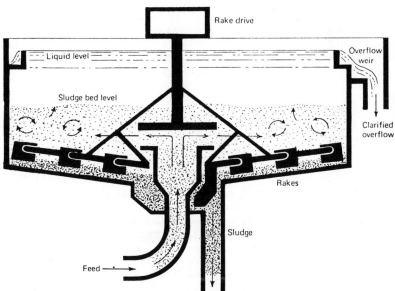

Figure 9.   Enviro-Clear thickener

initiated a coal preparation program commensurate with the vital
energy demands of this Nation. The former spectrum of coal preparation research has not been eliminated but rather broadened to
meet these new demands.

The primary objective of the Bureau of Mines Coal Preparation
Program is to increase the availability of environmentally acceptable fuels for combustion use through the removal of polluting
constituents. This is to be accomplished by advancing the state-
of-the-art of coal cleaning to a level where low-pollutant fuels
can be produced in a cost-effective manner in commercial-scale
quantities without altering the basic form of the processed fuel.

Two-Stage Coal-Pyrite Flotation. Froth flotation of coal
suffers as a desulfurization method because of the tendency for
the fine-size pyrite particles to float with the coal. As a con-
sequence, the Bureau of Mines has been conducting research to
improve flotation procedures so that better pyrite rejection is
possible. Early laboratory flotation work showed that the quantity of pyrite reporting with the clean coal froth product increased with increasing additions of the frothing agent and with
lengthy froth collection times. Therefore, some desulfurization
was made possible by using starvation amounts of reagent and very
brief slurry residence times.

Substantial sulfur reductions were also achieved with some
coals by two-stage rougher-cleaner flotation in which the clean
coal froth concentrate from the first stage was refloated. In all
of this earlier work, it was recognized that a portion of the
pyrite reported to the clean coal froth product because it was
attached to floatable coal or because it was too fine and became
mechanically entrapped in the froth. To combat this problem, a
unique two-stage flotation process was developed and patented (6)
by the Bureau. The process consists of a first-stage coal flotation step in which high-ash refuse and the coarse or shale-
associated pyrite are removed as tailings. The first-stage coal
froth concentrate is then repulped in fresh water, and a coal
depressant, a pyrite collector, and a frother are added. A high-
sulfur-content product is then floated, and the second-stage underflow is collected as the final clean coal. In laboratory and
pilot plant flotation tests with coals from various coalbeds
throughout the Eastern United States, pyritic sulfur reductions
of up to 90% were achieved using this technique (7).

The first commercial installation of the coal-pyrite flotation process will go on line at the Lancashire No. 25 preparation
plant of the Barnes and Tucker Co. in August 1977. This is a
cooperative effort with the Bureau of Mines. If it works suc-
cessfully, it will mean the recovery of an additional 8 tons per
hour of fine-size coal that is now being discarded as waste
because of its high sulfur content.

Research is being conducted as the University of Utah to
investigate the depressant and collector adsorption reactions

involved in the Coal-Pyrite Flotation Process.   The objectives of
this program will be to--
    (1)   Identify and characterize the hydrophilic
polymeric coal depressants.
    (2)   Determine the important operating variables
which control the adsorption-desorption reactions and
relate these results to the flotation response of the
mineral constituents.
    (3)   Establish procedures, if necessary, to allow
for subsequent coal flotation and plant water recycle
without deleterious effects from residual reagent
concentrations.

High-Gradient Magnetic Separation.   High-gradient magnetic
separation (HGMS) uses large-capacity devices as a practical
means to separate small, weakly magnetic particles on a large
scale.   This technology, which is being utilized commercially
to purify kaolin clay, is being investigated through several
Bureau of Mines contracts to determine its potential for application in the coal preparation industry.
    General Electric Co., in conjunction with the Massachusetts
Institute of Technology and Eastern Associated Coal Corp., is
attempting to establish the technical feasibility of removing a
substantial fraction of the inorganic sulfur from dry coal powders
by HGMS.
    In addition to sponsoring these contracts, the Bureau will
establish inhouse capability for HGMS in the near future.   A high-
gradient magnetic separator of the canister type has been ordered,
and a research program involving a wide variety of coals will be
initiated within several months.

Coal Prep/FGD Combination Study.   In a study just completed
for the Bureau of Mines by the Hoffman-Muntner Corp. (8), the
economic potential of coal preparation combined with stack gas
scrubbing was evaluated.
    The concept of physical coal cleaning combined with flue
gas desulfurization is not new.   In the Hoffman-Muntner study,
actual coal use areas, coal source areas, and the most probable
coalbed source are defined and used to make an economic evaluation of physical coal cleaning.   Then the cost of a new utility
plant removing SO_2 (exclusively) by stack gas scrubbing to meet
the new source emission standard was evaluated.   This was followed by a similar evaluation of the combined use of physical
coal cleaning plus stack gas scrubbing to attain the same SO_2
emission level.
    These case studies indicate that for many situations the
economic advantage of a combined approach is quite significant.
This advantage depends on:

(1) The availability of coals capable of significant
reductions in ash and sulfur by physical upgrading with
minimal Btu loss.

(2) A clean coal sulfur level that is compatible with
significantly less than full-scale scrubbing requirements.
Even so, the range of variables is such that each coal
source-user combination must be individually assessed.

Chemical Desulfurization. Present methods of physical coal
preparation in relation to sulfur reduction apply only to the
partial removal of pyritic sulfur. As these physical methods do
not remove organically bound sulfur, they can be applied effec-
tively only to selected coals. Therefore, the Bureau of Mines
program has been expanded to include research on experimental and
developmental techniques using chemical, in addition to physical,
methods for coal desulfurization so that more coals will be made
available within the framework of environmental restraints.

Presently, the Bureau is engaged in an inhouse research
program to develop innovative chemical treatments specifically
aimed at organic sulfur removal. The chemical oxidation of the
organic sulfur to a form which could be removed more readily
appears to have potential and is being investigated actively.

We envision that the optimum desulfurization of coal will
include both physical and chemical treatment. The raw coal will
be crushed fine enough to maximize pyrite release. It will then be
physically treated using wet processes to remove the liberated
pyrite. The fine-size wet coal product will then be treated by
a chemical process that will remove sufficient organic and remaining pyritic sulfur to provide an environmentally acceptable
final product.

Dewatering. Hot Surfactant Solution As Dewatering Aid (9).
A principal area of coal preparation research involves dewatering.
Moisture removal can be accomplished by mechanical methods, chemical treatment, or thermal drying. Coal finer than 28 mesh is the
most difficult size to dewater and is normally filtered and then,
when necessary, thermally dried.

Some early research work in our laboratory has suggested that
the use of surfactants during the vacuum filtration cycle will
increase the amount of water removed. The technique consists of
pouring a coal slurry onto a modified filter leaf apparatus and
washing the filter cake with a surfactant solution during the air-
drying cycle. The efficiency of the process depends on the amount
of surfactant in the wash water and on the temperature of the wash
water. It is possible to improve on the results by heating the
washed filter cakes with steam.

Using this technique, the moisture content of slurries made
with minus 35-mesh coal containing about 25% minus 325-mesh
material was reduced to about 10%. In contrast, untreated filter
cakes contained about 19% moisture.

Surface Phenomena in Coal Dewatering. An investigation to improve mechanical methods for dewatering fine coal to minimize the need for thermal drying is being carried out under a grant with Syracuse University.

Black Water Studies. Treatment of black waters from coal preparation plants is complicated by the heterogeneous nature of the waters and by the lack of information on the behavior of relatively simple systems. This water consists of mixtures of fine coal, clay minerals, quartz, calcite, pyrite, and other mineral particles dispersed in water. Effective treatment of the effluents must be carried out regardless of whether the water is to be reused or discharged. Under a grant, the Pennsylvania State University is investigating the treatment of black water from coal preparation plants by a flocculation process.

Refuse Pond Survey. In coal preparation plants utilizing wet cleaning methods, the fine coal refuse material constitutes about 10% of the total refuse material and is discharged in suspension either as tailings from the froth flotation process or as wet fines taken directly from the coarse coal washery circuits.

The University of Alabama currently has a grant from the Bureau to survey existing coal waste ponds in Alabama to determine the location of the wastes, the approximate quantity of coal in each waste pond, and the potential recovery of the coal based on washability tests of representative samples of the waste.

Refuse Pond Stabilization Study. Although refuse ponds continue to be used for tailings disposal, other considerations such as ground and surface water pollution, lack of suitable terrain, and environmental restrictions may render the ponding method totally unacceptable or prohibitively expensive in the future.

Under a contract just signed with the Dravo Lime Co., the critical parameters for landfill-type disposal will be established by determining the engineering properties and the physical and chemical characteristics affecting stabilization for typical samples of preparation plant wastes from the treatment of eastern bituminous coals. Calcilox additive, a Dravo-developed product, will be one of the principal stabilizing agents used in this study.

Lignite Upgrading. Lignite deposits in the North Dakota-Montana region represent one of the largest fossil fuel reserves in the United States. However, the utilization of this fuel has been limited by its high inherent moisture and its often high sodium content. Lignite contains about 38% moisture, and because of this high moisture content, the market area for lignite is limited to about a 300-mile range. Beyond this distance, the cost of delivered Btu's exceeds that of higher rank coals.

Reducing moisture content to 10% or less at reasonable costs
would make this low-sulfur coal competitive over a much wider
market area.

Other problems associated with the high moisture content of
lignite are freezing in railroad cars during cold weather and
slacking upon exposure to air, which causes dust pollution problems. Lignite is also extremely reactive toward oxygen and is
thus susceptible to spontaneous combustion.

The sodium content of lignite varies widely and may be as
high as 1.2%. Lignite containing less than 0.3% sodium can be
fired without difficulty while higher percentages of sodium
cause boiler tube fouling, reduced efficiencies, and excessive
downtime to remove ash deposits. The sodium content of
lignite can be reduced by ion exchange using metal salts which
supply cations that are exchangeable with sodium; their effectiveness depends upon the size of the lignite particle and the contact time.

The goal of our lignite-upgrading program is to produce a
final product containing less than 10% moisture and 0.3% sodium.
Crushed raw lignite will be processed in an ion-exchange vessel
where the sodium content will be reduced to an acceptable level.
The ion-exchange product will be dewatered, pelletized, and heat-
dried to the desired moisture content. This final product will
be a well compacted water-resistant pellet which will not combust
spontaneously and will have sufficient mechanical strength to
resist breakage during handling and storage.

Equipment Performance Studies. A significant part of the
Bureau's program for many years has been equipment performance
evaluations. These studies provide the necessary information for
coal preparation plant operators and equipment designers to
select and recommend existing coal washers. In this study,
only two types of coal-washing equipment remain to be evaluated:
jig washers, to be completed by the end of 1978, and froth flotation cells, to be concluded in 1981.

At this time, an evaluation of the performance of a Batac
jig is underway. Early in 1978, a study will begin covering the
principal pieces of coal dewatering and drying equipment. All of
these studies will be done by contract. Each is scheduled to
take approximately 3 years. Screens, contrifuges, vacuum and
pressure filters, and finally thermal dryers will be evaluated.

Trace Elements. There is an acute awareness that trace
elements in coal might contribute substantial quantities of
potentially hazardous materials to the environment. Since certain
trace elements concentrate selectively in particular specific
gravity fractions of raw coal, reduction of the mineral matter
may carry the concurrent benefit of reducing the trace element
concentration in the clean coal.

The Bureau of Mines recently completed a washability study
(10) showing the trace element contents of various specific gravity fractions for 10 coal samples collected from various U.S. coal-
producing regions. Reliable analytical methods were developed to
determine cadmium, chromium, copper, fluorine, mercury, manganese,
nickel, and lead in the whole coals and the various specific
gravity fractions of the coals.

Most of the trace elements of interest concentrated in the
heavier specific gravity fractions of the coal, indicating that
they are associated with mineral matter; removal of this material
would significantly reduce trace elements up to 88%.

In a related, but greatly expanded, effort, the Bureau,
through EPA, is about to fund a program with Bituminous Coal
Research. The objectives of this project are--

   (1) To prepare a comprehensive state-of-the-art
assessment of the fate of trace elements in coal mining,
preparation, transportation, and utilization; and
   (2) To determine the effect of coal cleaning on
trace elements.

Computer Simulation. To assist in the prediction of full-
scale preparation plant operation, a computer simulation program
is being developed under contract with the University of Pittsburgh (11). The standard prediction of product quality (such as
ash, sulfur, Btu, and yield) is calculated based on the extensive
equipment performance distribution data obtained from previous
U.S. Bureau of Mines investigations. Several other features,
though, are more novel and include the mathematical modeling of
the rotary breaker, other crushers, and both wet and dry screens.
The overall capability of the computer program is such that a
"build-your-own" preparation plant can be simulated.

Allied to this work is a study being conducted for the Bureau
by the Hoffman-Muntner Corporation. The information generated
under this contract will constitute the economic subroutine for
the total computer simulation program.

New Facility. In the United States there is no available
fully integrated demonstration-type facility for testing a new
coal washing technique, flowscheme, or piece of equipment. Consequently, a significant time lapse exists between development of
new equipment and techniques and their acceptance by the coal
industry.

To alleviate many of the past problems of introducing new
technology to the industry, a central coal preparation process
development facility has been designed. This will permit unbiased
engineering data to be scaled up readily to full-size commercial
coal preparation plant operation. Moreover, the expense of
evaluating processes that prove to be of limited value to the
industry will be reduced greatly.

The plant will have a nominal capacity of 10–25 tons per hour of raw coal, depending on the flowscheme used. Process flexibility is a prime design requisite. New crushing, screening, dewatering, and cleaning equipment and coal washing circuits will be evaluated. This plant will be used to process coals from different areas and will provide optimum operating conditions needed to produce a coal of acceptable sulfur, ash, and trace element content. An additional function of the demonstration plant will be to produce ton lots of specification coal for combustion testing, stack gas scrubbing work, or feedstock to coal conversion processes.

This brief overview of coal preparation research at the Federal Bureau of Mines shows a wide spectrum of interest in physical, as well as chemical, coal preparation. As the Nation gives increasing attention to energy-related problems, the pace of these coal preparation research efforts is likely to accelerate.

Literature Cited

1. Deurbrouck, A. W., Jacobsen, P. S., "Coal Cleaning—State-of the-Art," Coal and Environmental Technical Conference and Equipment Exposition, Louisville, Ky., Oct. 22-24, 1974.
2. Cavallaro, J. A., Johnston, M. T., Deurbrouck, A. W., "Sulfur Reduction Potential of the Coals of the United States," BuMines RI 8118, 1976, 323 pages.
3. Hucko, R. E., Cavallaro, J. A., "Effect of Coal Preparation on Northern Appalachian Reserves," 4th National Conference on Energy and the Environment, American Institute of Chemical Engineers, Cincinnati, Ohio, Oct. 3-7, 1976, pp. 307-316.
4. Killmeyer, R. P., Jr., "State-of-the-Art of Coal Cleaning and Recent Developments in Equipment and Circuits," Third Kentucky Coal Refuse Disposal and Utilization Seminar, Pineville, Ky., May 11-12, 1977.
5. Hucko, R. E., Deurbrouck, A. W., "Overview of the U.S. Bureau of Mines Coal Cleaning Program," Coal Contaminant/Removal Technology Conference, Research Triangle Park, N.C., Apr. 27-28, 1977.
6. Miller, K. J., "Flotation of Pyrite From Coal," U.S. Patent No. 3807557, Apr. 30, 1974.
7. Baker, A. F., Miller, K. J., Deurbrouck, A. W., "Desulfurization of Coal by Froth Flotation," Int. Coal Prep. Cong., Paris, France, March 1973, Paper 27E, 13 pp.; Available for consultation at the Bureau of Mines Coal Preparation and Analysis Laboratory, Bruceton, Pa.
8. Hoffman, L., Aresco, S. J., Holt, E. C., Jr., "Engineering/ Economic Analyses of Coal Preparation With SO$_2$ Cleanup Processes for Keeping Higher Sulfur Coals in the Energy Market," NCA/BCR Coal Conference and Expo III, Louisville, Ky., Oct. 19-21, 1976.

9. Baker, A. F., Deurbrouck, A. W., "Hot Surfactant Solution as Dewatering Aid During Filtration," 7th <u>Int. Coal Prep. Cong.</u>, Sydney, Australia, May 23-28, 1976.

10. Cavallaro, J. A., Gibbon, G. A., Hattman, E. A., Schultz, H., Deurbrouck, A. W., "A Washability and Analytical Evaluation of Potential Pollution From Trace Elements in Coal," Joint EPA/Bureau report, not yet published.

11. Gottfried, B. S., Jacobsen, P. S., "A General Distribution Curve for Characterizing the Performance of Coal Cleaning Equipment," BuMines RI 8238, to be published in 1977.

5

Chemical Comminution: A Process for Liberating the Mineral Matter from Coal

PHILIP H. HOWARD and RABINDER S. DATTA

Syracuse Research Corp., Syracuse, NY 13210

Conventional coal preparation consists of mechanical size reduction, which liberates the pyritic sulfur and other mineral matter, followed by a separation step, the cost of which depends upon the size consist of the crushed coal. In general, as the coal size consist decreases, the amount of mineral matter liberated and the cost of separation increase. Physical coal cleaning processes are not able to liberate organic sulfur, and, therefore, the organic sulfur concentration places an upper limit on the amount of sulfur that can be removed.

Chemical comminution provides a unique way of crushing coal for mineral matter liberation. Instead of mechanical size reduction, the coal is treated with a chemical (usually ammonia gas or a concentrated aqueous ammonia solution), resulting in selective breakage which appears to occur along the bedding planes and along the mineral matter (e.g., pyrite) and maceral boundaries. Washability comparisons of mechanically crushed and chemically comminuted coal samples have indicated that, at a given size consist, more liberation of pyritic sulfur and comparable liberation of ash are possible with chemical comminution. Figures 1, 2, and 3 provide a comparison of typical mechanical and chemical breakage and liberation. The recoveries noted in Figures 2 and 3 only apply to +100 mesh product, and, therefore, the −100 mesh material should be considered when calculating recoveries based upon feed. In terms of decreasing size consist (Figure 1), the following order is found: 1 1/2" top size ROM > chemically fractured > 3/8" top size mechanically crushed > 14 mesh top size mechanically crushed. The same order is found for the ash vs. recovery curves in Figure 2. In contrast, the total sulfur vs. recovery curves (Figure 3) demonstrate that chemical fracture (only 4.53% is < 100 mesh) liberates considerably more pyritic sulfur than mechanical crushing even to −14 mesh (21.9% is < 100 mesh). Similar results have been found with Redstone, Pittsburgh, and Upper Freeport seam coals (1,2,3) and with some Iowa coals (4). Keller and Smith (5) have reported that greater sulfur liberation does not occur

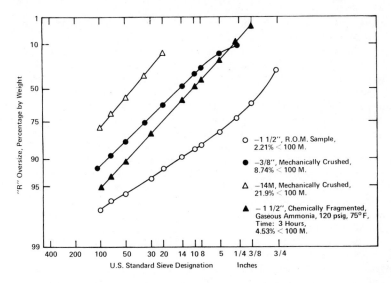

Figure 1. Size consist of Illinois No. 6 coal samples (Franklin County, IL)

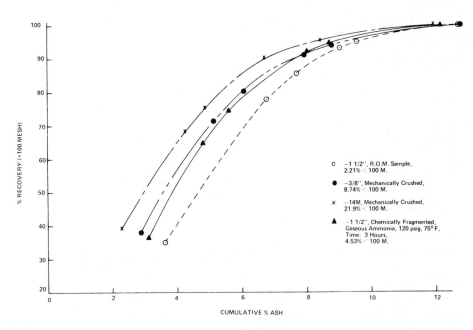

Figure 2. Ash vs. recovery curves comparing the 3-hr chemical comminution to the mechanical crushing of the Illinois No. 6 seam coal sample (Franklin County, IL)

with chemical fracturing but they did not consider yields in their comparisons (6).

The fact that chemical comminution can liberate more of the pyritic sulfur without grinding to small sizes has considerable economic benefits. Although, at this stage in the development, it is difficult to estimate the exact costs of a coal preparation flowsheet using chemical communition as the size reduction method, some preliminary estimates have been developed (3). The total capital and operating costs for the chemical treatment alone using ammonia vapor, under conditions shown to be technically feasible in the laboratory, vary from $1.00 to $1.50 per ton of coal product. Using relatively inexpensive density separation techniques, feasible because of the small amount of fines produced in the chemical fragmentation step, would bring the total cost for producing clean coal to between $2.50 - $3.00 per ton of product. This is very competitive with other processes currently being considered for producing clean coal. It is envisaged that the chemically comminuted product after cleaning will contain 80-90% less pyritic sulfur and 50-60% less ash, and, in approximately 30-40% of the Northern Appalachian seam coals, will meet EPA new source emission standards.

Because of the commercial significance of chemical comminution, the effects of different reaction conditions have been studied preliminarily to provide some insight into the mechanism of the phenomena. The available information is presented in the following sections, and further information is being developed.

Equipment and Procedures

The procedures used to chemically comminute 100 lbs. samples (Figures 1, 2, and 3) are described elsewhere (3). Smaller samples were comminuted in the apparatus depicted in Figure 4. Unless indicated, the smaller coal samples were floated at 1.6 specific gravity to remove large pieces of mineral matter and then closely sized (e.g., 3/4" x 1/4") so that the fragmentation rates could be compared. The operation of small bomb apparatus is outlined in Procedure A.

The fragmentation rates under different conditions or with different coals were compared in two ways. One way was to compare the weight percent of the various coal samples that passed 5 mesh. The other method consisted of plotting the screen analysis of the coal samples on a Rosin Rammler graph and comparing the absolute size constant $(\bar{x})$. The absolute size constant is the size where 36.79% of the coal particles are retained on the screen. Using the absolute size constant for comparison requires that the slope of the size distribution remain approximately constant. With chemically comminuted samples, a constant slope has been observed.

Mineral Liberation

Microscopic examination of chemically comminuted coal has been conducted by Greer (7) of Iowa State University using a scanning electron microscope. The results demonstrated that fragmentation from chemical treatment was strongly controlled by maceral boundaries and other deposits within the material such as pyrite bands.

The above result demonstrates the selective breakage that occurs with chemical comminution and explains why pyrite is liberated during chemical treatment without excessive size reduction. However, the difference between sulfur and ash liberation (Figures 2 and 3) has not been determined. Further petrographic studies of this effect are anticipated.

Effective Chemicals

Although a number of chemicals have some comminution ability (8), the chemicals that appear to have the greatest effect are ammonia (gas and anhydrous and hydrous liquid) and methanol. These compounds fall in a class of chemicals containing a nonbonding pair of electrons (oxygen and nitrogen compounds) which swell (9,10) and dissolve (10) coal at ambient temperatures. Although swelling studies have not been conducted by us, very little coal (< 0.1%) was dissolved by either methanol or liquid anhydrous ammonia. The swelling effect, which has been observed with methanol-treated coal by Bangham and Maggs (9), may cause the fragmentation which occurs during chemical treatment. Keller and Smith (5) have suggested that solvent swelling is the mechanism causing the breakage. However, their only support of this contention was a theoretical discussion. The fact that coal does comminute with gaseous ammonia is not explained by this theory (6). Another analogy between coal solvents and coal comminutants is a reduction in effect as the solvent is diluted with water. Other specific solvents mentioned as good coal solvents (10), such as n-propylamine and pyridine, have been briefly examined. These chemicals do cause fragmentation but are not as effective as ammonia. Since these chemicals are larger in molecular size, it is possible that molecular size is an important parameter for chemical comminution, especially if penetration of the coal is a rate-determining factor.

Effect of Reaction Conditions

The fragmentation caused by chemical treatment is affected by such parameters as moisture, pressure, water concentration in the chemical, starting size of the coal, and preconditioning of the coal before treatment. These effects, using Illinois No. 6 seam coal as an example, are illustrated in Figures 5, 6, 7, and

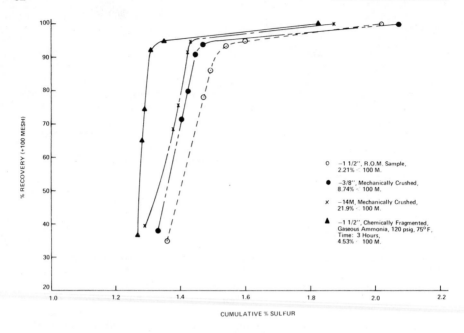

Figure 3. Sulfur vs. recovery curves comparing the 3-hr chemical comminution to the mechanical crushing of the Illinois No. 6 coal sample (Franklin County, IL)

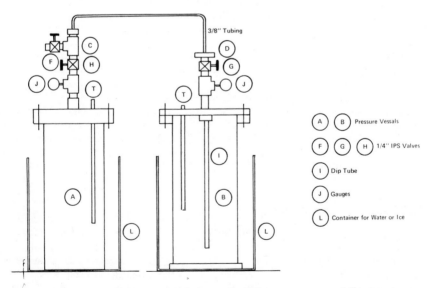

Figure 4. Small bomb apparatus. Procedure given on following page.

Procedure A

Procedure followed when performing tests in small pressure vessels depicted in Figure 4.

At Start: Covers removed from Bombs (A) and (B) .
            Piping on covers in place up to unions (C) and (D) .
            Dip tube (I) in place.

Step No.

1	Weigh 300 gms. coal into (A) .
2	Close (A) with cover and attach piping to (C) .
3	Evacuate (A) via connection at (C) .
4	Close valve (F) .
5	Cool down (B) with dry ice around it.
6	Add hydrous or anhydrous NH_3 to near 1/2 full.
7	Close cover with attached piping, valve (G) open, and tighten nuts.
8	Close valve (G) .
9	Install connecting tubing between (C) and (D) .
10	Evacuate tubing through valve (F) .
11	Remove dry ice from round (B) and add warm water to containers (L) so as to maintain temperature at 75°F.
12	Allow pressure to build up in (B) to a constant amount, maintaining water jacket temperature at 75°F.
13	Open valves (H) and (G) , forcing contents from (B) to (A) until pressure in (A) and (B) equilizes.
14	Start timer (recording temperature and pressure), maintain temperature of water jacket at 75°F. Close valve (H) .
15	Disconnect (A) from tubing at (C) .
16	60 sec. before the duration of the experiment, remove water jacket from round (A) and cool (A) down with dry ice.
17	Vent (A) to 0 p.s.i.g. through valves (H) and (F) .
18	Remove cover and separate liquid from solid.
19	Remove the residual NH_3 in coal by ellutriating the sample with boiling water in an ellutriating column.
20	Collect sample, dry, and perform appropriate analyses.

Note: 1. In case of NH_3 gas treatment, the dip tube (I) has to be removed.

        2. For making hydrous NH_3, a known volume of anhydrous NH_3 is added to a known weight of ice.

8. Illinois No. 6 seam coal is very susceptible to chemical
comminution, and, therefore, the results with this coal are not
necessarily representative of other coals. Figure 5 demonstrates
the importance of evacuating the reactor before chemical
treatment. The contrast in the effect of evacuation between
liquid and gaseous conditions is quite apparent. This effect
has also been noticed with other coals, although they have not
been as demonstrative. For example, all the conditions used in
Figure 5 would have no fracturing effect on a Pittsburgh seam
coal that was examined.

Figure 6 illustrates the effect of pressure and water
content and demonstrates that methanol is not as effective a
comminuting agent as even a dilute ammonia-water solution. With
gas treatment, it appears that a little moisture in the coal
aids fracture. Also, when using gas, a change of pressure from
90 psig to 120 psig has considerable impact on the amount of
breakage. Determination of any trends with the liquid systems
is difficult because the pressure was not held constant. As
might be expected, the initial size of the coal before treatment
can effect the size of the treated product. This effect is
illustrated in Figure 7.

As noted previously (Figure 5), evacuation before treatment
appears to have a considerable effect, especially with gaseous
treatment. Evacuation after treatment also appears to have an
effect (Figure 8) which may be caused by just a difference in
reaction time, since reaction in the evacuated sample should
stop rapidly while the unevacuated sample may continue to react
even after the pressure is removed.

Effect of Coal Type

The reaction conditions have a considerable impact on the
amount of breakage that occurs with different coal ranks
(Figure 9). Gaseous ammonia is least affected by rank while
liquid ammonia either at elevated or at atmospheric pressure
is strongly affected. The liquid ammonia correlation between
rank and comminution may be fortuitous and caused instead by
differences in micro or macro porosity, maceral content, cleat
system, swelling ability (11), mineral matter distribution, or
perhaps other factors.

Chemical Reactions

Chemical reactions between the ammonia and coal could have
an adverse effect on the recovery of the ammonia and on the
amount of nitrogen oxides emitted when the coal is combusted.
Therefore, the nitrogen content of coal before and after ammonia
treatment was determined for a variety of coal seams. The
results presented in Table I vary slightly for different coals,
and the increase in nitrogen appears to be in correlation with a

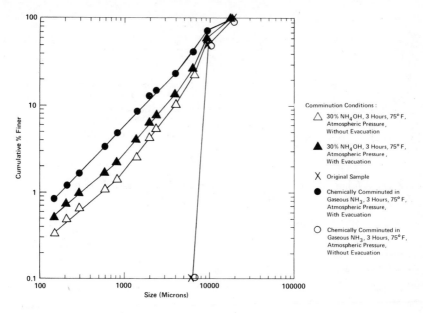

Figure 5.   Effect of preevacuation on chemical comminution of Illinois No. 6 coal

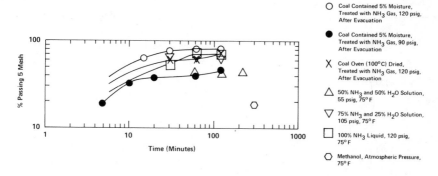

Figure 6.   Effect of reaction pressure and water content in the coal or comminuting agent on fragmentation of ¾ in. × ¼ in. samples of Illinois No. 6 coal.

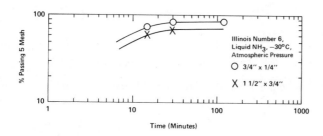

Figure 7.   Effect of starting size on chemical comminution of Illinois No. 6 seam coal

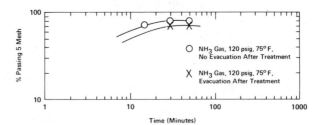

Figure 8.   *Effect of evacuation after chemical treatment of Illinois No. 6 seam coal*

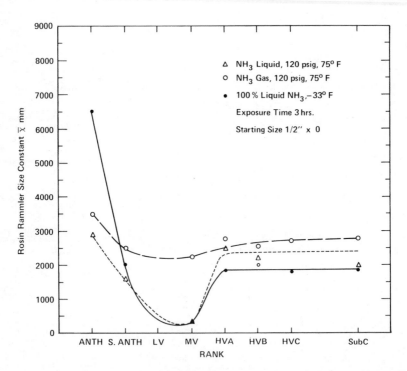

Figure 9.   *Effect of coal rank on the amount of breakage at various reaction conditions*

Table I. Nitrogen Content of Chemically Comminuted Coal[a]

Coal	Ammonia Treatment	Treatment to Remove Ammonia	Original ROM Sample	Air Dried at 60°C Overnight	Air Dried at 60°C and Sized				Air Dried at 100°C for 4 hr	Air Dried at 200°C for 2 hr	Air Dried at 300°C for 1 hr
					+8M	8x20M	20x100M	-100M			
Pittsburgh, Green County, PA	180 min, NH$_3$ gas, 120 psig, 75°F	air dried, 60°C	1.43	1.51							
	30 min, 100% NH$_3$ liquid, 120 psig, 75°F	rinsed with dilute HCl at 20°C, air dried		1.53							
		elutriated with water at 100°C, air dried		1.50							
Upper Freeport, Westmoreland County, PA	240 min, 100% NH$_3$ liquid, atm. press., -30°F	air dried	1.21	1.12	1.08	1.20	1.44	1.21	1.21	1.20	1.24
		elutriated with water 100°C for 1/2 hr		1.21	0.97	1.15	1.37	1.22	1.24	1.37	1.38
		elutriated with water 100°C for 2 1/2 hr		1.25	0.98	1.21	1.39	1.16	1.24	1.26	1.11
Illinois No. 6, Franklin County, IL	240 min, 100% NH$_3$ liquid, atm. press., -30°F	air dried	1.58	1.94	1.75	1.96	1.94	1.64	1.91	1.94	2.00
		elutriated with water at 100°C for 1/2 hr		1.80	1.58	1.74	1.69	1.59	1.68	1.69	1.82
		elutriated with water at 100°C for 2 1/2 hr		1.72	1.62	1.74	1.71	1.43	1.70	1.69	1.96

[a] Numbers in percent total nitrogen.

decrease in rank. No increase appears to take place with Upper
Freeport seam coal, a slight increase (6%) with Pittsburgh seam
coal, and approximately a 20% increase with Illinois No. 6 seam
coal when the sample is air dried. However, some of the nitrogen
can be removed by hot water washing. In general, the +8 mesh
particles show a lower increase in nitrogen than the other
sizes. The nature of the chemical reaction that may be taking
place is unknown, but there are functional groups (e.g., esters)
in coal that could form nitrogen compounds. With Illinois No. 6
seam coal, the loss of ammonia would still be small (4.4 lbs.
ammonia per ton of treated coal when a hot water wash is used),
and it is unknown whether nitrogen oxides emissions would change.

From the above results, it can be seen that considerably
more information is necessary before the mechanism of chemical
comminution is understood. Studies directed at a better
understanding of the phenomena and the effect it has on the
chemically treated coal are presently underway.

Acknowledgements

This work was partially supported by ERDA Contract
14-32-0001-1777.

Literature Cited

1. Datta, R.S., Howard, P.H., Hanchett, A., "Pre-Combustion
    Coal Cleaning Using Chemical Comminution," NCA/BCR
    Coal Conference and Expo III. Coal: Energy for
    Independence, Louisville, October 19-21, 1976.
2. Howard, P.H., Hanchett, A., Aldrich, R.G., "Chemical
    Comminution for Cleaning Bituminous Coal," Clean Fuels
    from Coal Symposium II Papers, Institute of Gas
    Technology, Chicago, Ill., June 23-27, 1975.
3. Datta, R.S., Howard, P.H., Hanchett, A., "Feasibility Study
    of Pre-combustion Coal Cleaning Using Chemical Comminution,"
    FE-1777-4, Final Report ERDA Contract 14-32-0001-1777,
    Syracuse Research Corp., November 1976.
4. Min, S., Wheelock, T.D., "Cleaning High Sulfur Coal,"
    NCA/BCR Coal Conference and Expo III. Coal: Energy for
    Independence, Louisville, October 19-21, 1976.
5. Keller, D.V., Jr., Smith, C.D., "Spontaneous Fracture of
    Coal," Fuel (1976), 55, 273-280.
6. Howard, P.H., Datta, R.S., Hanchett, A., "Chemical Fracture
    of Coal for Sulfur Liberation," Fuel (1977), 56, 346.
7. Greer, R.T., "Coal Microstructure and the Significance of
    Pyrite Inclusions," Scanning Electr. Microscopy, Vol. I,
    Proceedings of the Workshop on Materials and Component
    Characterization/Quality Control with the SEM/STEM,
    IIT Research Institute, Chicago, Ill., March 1977.

8. Aldrich, R.G., Keller, D.V., Jr., Sawyer, R.G., "Chemical
    Comminution and Mining of Coal," U.S. Patent 3,815,826
    (June 11, 1974) and 3,850,477 (November 26, 1974).
9. Francis, W., "Physical Considerations," Chapter XI in
    "Coal, Its Formation and Composition," 2nd ed., Edward
    Arnold, London, 1961.
10. Dryden, I.G.C., "Solvent Power for Coals at Room
    Temperature," Chem. Ind. (June 2, 1952), 502.
11. van Krevelen, D.W., "Chemical Structure and Properties of
    Coal. XVIII-Coal Constitution and Solvent Extraction," Fuel
    (1965), 44, 229.

6

Use of the Flotation Process for Desulfurization of Coal

F. F. APLAN

Mineral Processing Section, Pennsylvania State University, University Park, PA 16802

The flotation process is a method for desulfurizing fine coal whose time has come. The process has proven to be a great success in the beneficiation of a wide variety of metalliferous ores and industrial minerals, and currently, the United States has an installed capacity to treat about 1.7 million tons of ore per day. Unfortunately, coal processing has not participated in this growth in any major way. In the U.S. in 1973 only 14 million tons of clean coal were produced by the flotation process (1). By comparison, in the same year approximately 400 million tons of coal were treated in preparation plants to produce 300 million tons of clean coal. This amount of coal produced by washing techniques was about half of the 600 million tons of the marketable coal produced in the U.S. that year.

There are many reasons why coal flotation has not made a greater impact on the coal industry, but chief among them are the effectiveness of coarse cleaning processes (which treat particles of 1-5 in. top size), the low value of a ton of coal until recent years, the lack of an adequate research and development program to evaluate the process thoroughly, and, to some extent, inertia. To these reasons should be added the following shortcomings of the coal flotation process enumerated by Mitchell (2) in 1948:

(1) Marketing problems with the fines,
(2) Cost of dewatering,
(3) Indifferent test results in removing sulfur,
(4) Inability to make clean separations with the finer sizes,
(5) Inability to clean slurries containing a high percentage of clay.

Two new factors, which have come to the fore within the past decade, have caused a serious re-evaluation of the process in coal preparation: the sharply increased market value of a ton of coal and environmental considerations. The increased coal price has led to a greatly increased emphasis on fine coal cleaning in order to achieve a greater yield of combustible material from a ton of raw coal. On the environmental front coal flotation helps to clarify the large quantities of water used and recycled in the

preparation process and in the removal of pyritic sulfur from the
coal. While for coarse coal, gravity concentration processes are
very efficient, their fine cleaning counterparts are not nearly
as effective in either coal recovery or in pyrite removal. The
coal flotation process offers the potential of overcoming both of
these defects for the treatment of the fine, -28 mesh coal. In
this respect it can reduce the sulfur content of some coals to EPA-
acceptable levels or, for higher sulfur coals, can reduce the overall sulfur content of the coal such that the need for flue gas
desulfurization at the power plant can be minimized. Where
applicable, coal preparation techniques can remove pyritic sulfur
for a fraction of the cost of sulfur removal by a flue gas
scrubbing system. In a recent article Hoffman et al. (3), have
shown the economic potential of such combination coal preparation
and flue gas desulfurization techniques.

The coal flotation process has been detailed in the literature
in review articles by Aplan (4), Brown (5), and Zimmerman (6).
Unfortunately, since the historical purpose of coal preparation
has been to remove the high-ash constituents, the literature does
not contain a great deal of information on sulfur reduction by
flotation. For this reason, the Mineral Processing Section at The
Pennsylvania State University several years ago embarked on a
comprehensive program aimed at understanding the fundamentals of
the process, delineating the best conditions for rejecting
pyritic sulfur during coal flotation, and establishing a data base.
This chapter will emphasize that work.

General Methods of Pyrite Removal

There are three general methods that may be used to reject
pyritic sulfur during coal flotation: (1) multiple cleaning,
(2) pyrite depression, and (3) coal depression with concurrent
flotation of the pyrite.

The first process is based on the assumption that coal is
easily floatable whereas pyrite is not. In this case any pyrite
which floats would be caused by mechanical entrapment with the
floating coal. It could thus be removed by repulping the floated
coal and repeating the froth flotation process several times; each
time rejecting some of the mechanically entrapped pyrite particles.
The assumption that pyrite is not floatable is true only as a
gross generalization, since, in practice, fine pyritic sulfur
often floats with the coal. The flotation of pyrite occurs not
only by itself, but its flotation is often encouraged by the
existence of locked pyrite-coal particles or by the use of an oily
collector for the flotation of the coal. Early-day ore flotation
processes, circa 1905, used fuel oil in the flotation of sulfide
minerals including pyrite. Studies by Miller et al. (7) and more
recently by Im (8) show that the multiple cleaning method is not
only laborious but often not particularly effective. The froth
sprinkling method, also evaluated by Miller (9) is a mutation of

the multiple cleaning process.

Pyrite depression during coal flotation would appear to be the most logical approach to the problem especially in view of the fact that pyrite depression techniques during the flotation of copper, lead, nickel, and zinc sulfides are well known (10). In coal flotation use has been made of pH control, lime, sodium cyanide, sodium sulfide, hydrosulfide and sulfite, potassium dichromate, potassium permanganate, and ferrous and ferric sulfate to reject pyrite (4). The classic work in this area is that of Yancey and Taylor (11), and more recent work has been done by Miller (12). Zimmerman (13) demonstrated that the sulfur content of clean coal produced by flotation of Pittsburgh Seam coal decreases as the pH is raised from 4 to 11. Unfortunately, the recovery of clean coal decreases as the pH is increased above pH 7. The practicality of the situation is that none of these techniques are used industrially, and laboratory testing often shows mediocre results at best. Truly, this is an area where an expanded research and development effort is clearly needed.

The third method of pyrite rejection is that of coal depression and has recently been tested by the U.S. Bureau of Mines (14) in a process called the two-stage or reverse flotation process. It involves a first-stage (rougher) flotation of the coal to reject the bulk of the ash-containing constituents including much of the pyrite. The rougher froth is then sent to a cleaner circuit where a proprietary depressant (American Cyanamid Reagent 633) is added to depress the coal while a xanthate collector and a frother are added to float pyrite from the coal. The process has rather strikingly reduced the sulfur level of the clean coal although the problems of excessive reagent use and cost and inadequate recovery of clean coal will require additional study. There is a need to evaluate coal depressing agents, and these studies are presently in progress (8).

Response of Pyrite in Coal Flotation Systems

Because of the dearth of data concerning the response of pyrite in a coal flotation circuit, the current program at Penn State has focused on establishing a proper data base. Factors to be considered in addition to the nature of the coal are the amount and kind of frother, quantity and type of collector (if any), type of flotation cell, and operating conditions. The following report on the experimental program in progress evaluates many of these factors.

Experimental Methods. All of the work reported here, except that in the first figure, has been done with clean coal from the Lower Kittanning or Pittsburgh Seam to which 5% purified pyrite has been added. Recent work has shown (15) that pyrite samples from different coal sources possess different floatabilities (which, in turn, are different from the floatability of ore pyrite).

Therefore, the pyrite used in each test came from the same seam as did the coal. This procedure eliminates locked particles so that the path of the pyrite during flotation can be accurately traced, uncomplicated by the intermediate response of particles that are part coal and part pyrite. Other studies are underway to evaluate the response of locked particles in coal flotation systems.

Coal from the Lower Kittanning seam (Cambria County, PA.) or the Pittsburgh seam (Fayette County, PA.) was pulverized to 100% -14 mesh, 90% -28 mesh, to which -28 mesh purified pyrite from each respective seam was added. Coarser pyrite was not added since Rastogi and Aplan (16) have demonstrated that +28 mesh pyrite will not float under the conditions normally encountered in froth flotation. To insure that surface oxidation did not influence the results, the coal and pyrite samples were ground shortly before flotation. The older style Fagergren laboratory flotation cell was used, and the pulp concentration was about 17% solids. The specific data given in this report were drawn from studies on two different coal seams and thus are intended to illustrate general principles rather than to serve as a basis for a flotation scheme for a given coal sample.

In using the data, one caution should be noted: since 95% clean coal and 5% pyrite were used, the maximum yield of combustible material is 95%. Yields approaching and exceeding this value are most certainly accompanied by the flotation of much liberated pyritic sulfur (SLP).

Coal Flotation Kinetics. Historically, the results of soft coal preparation have been most often analyzed in terms of yield and of product ash and, more recently, sulfur. As a result of our extensive studies on the nature of coal flotation, however, we believe that the rate of coal flotation should be added to the list of evaluation criteria used for laboratory studies. It has been amply demonstrated (16) that different fractions in the raw coal float at different rates, and thus the coal flotation rate is a variable that can be used to achieve better separations, most particularly between coal and pyrite.

Coal flotation follows a first-order rate equation (16):

$$-\frac{dC}{dt} = k\ C \tag{1}$$

where C = concentration of floatable material; t = time, sec; k = rate constant, sec^{-1}.

This agrees with the findings of Bushell for the flotation of quartz with an amine (17) and the findings of numerous other investigations for ores (18). The experimental approach was to use a batch flotation method and to apply the analysis proposed by Bushell (17). Figure 1 for the flotation of Pittsburgh Seam coal with MIBC, shows that coal flotation follows the curve AB_3C_3, and this curve is a straight line, and hence follows Equation 1, over the segment $A-B_3$. This corresponds to the flotation of 80% of the

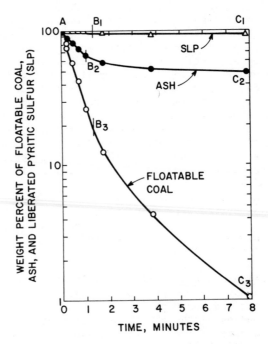

American Institute of Mining,
Metallurgical and Petroleum Engineers

Figure 1. The rate of flotation of Pittsburgh seam coal using 0.84 lb/ton pine oil. Percentage remaining in the flotation cell as a function of time (16).

coal, and, in fact, there is only slight deviation from linearity for the flotation of 90% of the coal.

The first-order rate constant, in sec^{-1}, may be determined graphically from plots such as those given in Figure 1 or may be calculated over the straight line portion of the curve by the following equation:

$$k = 2.303 \frac{\log C_1 - \log C_2}{(t_2 - t_1)} \quad (2)$$

where C_1 = concentration of material at t_1; C_2 = concentration of material at t_2.

From Figure 1 it may be seen that the flotation of the ash-bearing constituents and the liberated pyritic sulfur (SLP) also follows the first-order rate law over the first minute to minute and one-half of flotation. This period corresponds to the flotation of nearly all of the coal present. Numerous other flotation tests with a variety of coal samples, frothers, collectors, particle sizes, and operating conditions confirm the accuracy of this interpretation (15,16,19,20). The slow floating fraction – that floating after $\sim$ 90 sec – is composed of coarser particles, those whose surface is slightly oxidized, lower rank and high ash fractions such as fusinite, and, in raw coal systems, locked particles. As will be demonstrated later, kinetic analysis is a valuable aid in evaluating the complicated interactions between coal type, frother system, and flotation operating conditions.

Effect of Frother Systems. A comparison of frothers is given in Table I (15) showing the yield (after 7.75 min of flotation time), percentage of liberated pyritic sulfur (% SLP), and the amount of coarse, 14 x 28 mesh, coal floated. An equimolar concentration of each frother was chosen as the basis of comparison.

TABLE I. Frother Comparison for the Flotation of Lower Kittanning and Pittsburgh Seam Coals (15).

Coal Type	Frother	Concentration		Yield,[a] %	Floated mat'l. %	
		mol L.	lb. ton		SLP[b]	+28 mesh
Lower	MIBC[c]	2.4×10^{-4}	0.28	94	20	92
Kittanning	Pine Oil	2.4×10^{-4}	0.44	96	33	91
	Cresol	2.4×10^{-4}	0.29	69	--	24
Pittsburgh	MIBC[c]	2.4×10^{-4}	0.28	84	16	29
Seam	Pine Oil	2.4×10^{-4}	0.44	89	21	48
	Cresol	2.4×10^{-4}	0.29	19	2	2

[a] Coal yield after 7.75 min of flotation.
[b] SLP = Liberated pyritic sulfur.
[c] MIBC = Methyl isobutyl carbinol, 4-methyl-2-pentanol.

From this table several conclusions emerge:
   (1) MIBC and pine oil are roughly equivalent frothers on an
equimolar basis, but MIBC is superior on a pounds per ton basis
   (2) Cresol is a much inferior frother to either MIBC or pine
oil
   (3) Pine oil tends to float slightly more pyrite than does
MIBC
   (4) Pine oil is a superior frother for the flotation of
coarser coal particles, especially for an intermediate floating
coal such as the Pittsburgh seam
   (5) The Lower Kittanning coal is a faster floating coal than
is that from the Pittsburgh seam
   These are only general conclusions, and the nature of the
coal and the nature and concentration of the frother may well
change the order of things. Most critical is the problem of
impurities which tend to float pyrite. The problem of impure
cresols is particularly critical; the oily impurities act as
collectors and, hence, favor the flotation of coarse and other
difficult-to-float particles--but at the expense of encouraging
the flotation of undesirable pyrite.
   Oil has long been used to enhance the flotation of coal
particles otherwise difficult to float with a frother only. The
introduction of an oily collector is, however, accompanied by the
flotation of much undesirable pyrite. This is clearly illustrated
in Table II (21) where the introduction of 0.46 lb./ton of fuel

TABLE II.   Effect of Fuel Oil on Increasing the Flotation of
            Pyritic Sulfur During the Flotation of Lower Kittanning
            Coal (21).

MIBC lb./ton	Fuel Oil lb./ton	Flotation rate x 10^3, sec^{-1}			Yield,[a] %	SLP floated, %
		k_C[b]	k_{SLP}	k_C/k_{SLP}		
0.17	0	37.5	2.3	16	90	13
0.17	0.46	55	12.4	4	97	56

[a] Coal yield after 7.75 min of flotation.
[b] k_C = Flotation rate constant for coal.

oil to a coal flotation system using MIBC as frother increased the
amount of the liberated pyritic sulfur (SLP) which floated from
13 to 56%!!! In this case the kinetic analysis is helpful.
Numerous other data (15,16,19) have established that coal generally
floats about 10-30 times faster than does pyrite; the exact value
of the ratio depends on the specific coal and pyrite and on the
frother system and flotation operating conditions. An important
evaluation criterion is thus to look at the ratio of the flotation
rate constant of coal to that of pyrite, k_C/k_{SLP}. As the value of

this ratio differs from an average value of, say, 16, those
flotation conditions either favorable or unfavorable for pyrite
rejection may be readily discerned. It can be seen from Table II
that with the addition of fuel oil, the coal flotation rate
increased somewhat but that the pyrite flotation rate increased
five-fold; the k_C/k_{SLP} ratio dropped to 4. Oil encourages the
flotation of pyrite much more than it does that of coal in this
instance.

Effect of Frother Concentration. Figures 2 and 3 show the
effect of an increasing amount of the frother, pine oil, on coal
yield and the amount of liberated pyritic sulfur (SLP) floated as
a function of flotation time (22). With respect to yield (Figure
2), there is a threshold amount of frother needed to obtain good
coal yields. Beyond this incipient amount of frother, however,
additional frother does not further increase yield with this coal.
The same pattern holds for a flotation time of 0.25 or of 7.75 min.
A 1-min flotation time is sufficient to achieve nearly complete
yield for frother concentrations of 0.5 lb./ton or more. However,
with the flotation of pyrite (Figure 3), a different pattern
emerges: both an excess frother dosage and extended flotation
times lead to the flotation of an ever-increasing amount of pyrite.
The advantage of using short flotation times and a low frother
concentration is obvious.

Effect of Flotation Operating Conditions. Table III compares
three levels of air rate and flotation cell impeller speed (16).

TABLE III. Effect of Air Flow Rate and Impeller Speed on the
Flotation of Pittsburgh Seam Coal (16).

Cell Operating Conditions[a]		Flot. rate x 10^3, sec^{-1}			Flotation Data[d]		
air rate (cfm)	speed (rpm)	k_C[b]	k_{SLP}[c]	$\dfrac{k_C}{k_{SLP}}$	Yield	% SLP[c] floated	% + 28M floated
0.535	2540	73	4.5	15.9	92.4	27.9	70.3
0.233	1660	25.5	---	---	79.1	17.7	29.8
0.136	1380	15.3	0.2	76	67.8	1.4	15.0

[a] MIBC 0.57 lb./ton.
[b] k_C = Flotation rate constant for coal.
[c] SLP = Liberated pyritic sulfur
[d] One minute flotation time.

Intense flotation conditions (high aeration rates and high impeller
speed) favor the fast flotation of coal, the flotation of coarse
particles, and the flotation of pyrite. Again, using the ratio of
the flotation rate constants for coal and pyrite, k_C/k_{SLP}, the
maximum rejection of pyrite occurs under gentle operating

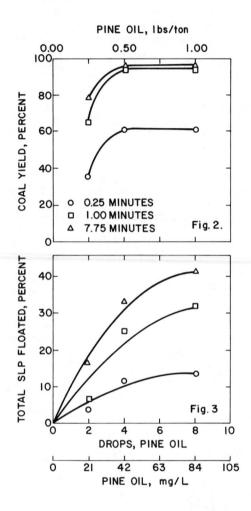

American Institute of Mining,
Metallurgical and Petroleum Engineers

*Figure 2. (top) Flotation yield of Lower
Kittanning coal as a function of pine oil concentration and time (22)*

American Institute of Mining,
Metallurgical and Petroleum Engineers

*Figure 3. (bottom) Effect of pine oil dosage
on the amount of liberated pyritic sulfur
floated. Lower Kittanning coal (22).*

conditions. Long flotation residence times favor the flotation of both coarse particles and pyrite.

Rastogi and Aplan (4,16) have demonstrated that the air rate alone is responsible for the bulk of the increase in coal flotation. The contribution of impeller speed is modest.

Test Evaluation. A study of numerous flotation tests shows that there is a relationship between yield and the amount of pyrite which floats (Figure 4) (4,23). This curve is similar to the grade-recovery curves often used in ore dressing studies (24) and to coal-sulfur data reported by Miller (12). Each particular flotation system shows a similar family of curves. Ash, total sulfur, or flotation rate may also be used as the abscissa and similarly shaped curves result (4,23). This method may be used to save considerable time in flotation testing since sulfur values at intermediate yields may be readily estimated without the need to resort to still another test. The bulk of the free pyrite that floats accompanies the last 20% or so of yield. This fact has been used to develop a coal flotation procedure called the "grab and run" technique.

The "Grab and Run" Process. In this method approximately one-half to two-thirds of the ultimate coal yield is floated in the rougher cell under gentle operating conditions (starvation quantities of a gentle frother such as MIBC, low air rate, low impeller speed, short flotation residence time, etc.) to remove a fraction low in pyrite. Under most conditions studied so far, this fast-floating fraction can go directly to the clean coal product; hence the name "grab and run." The balance of the recoverable coal is removed in a scavenger circuit under more intense operating conditions. As the ultimate yield is approached, the amount of pyritic sulfur that floats increases greatly. The scavenger concentrate, which is higher in pyritic sulfur than the rougher concentrate, is sent to cleaner flotation. Here two options are available: depress pyrite and float coal or vice versa. In the latter option the scavenger cleaner circuit resembles the U.S. Bureau of Mines reverse flotation process except that only a fraction of the total tonnage need be treated. Details of this process will be reported soon (25).

Summary. Pyrite is best rejected under gentle frothing and operating conditions using short residence times in the flotation cell. The use of excess frother and, especially an oily frother and/or collector, is particularly counter-productive to pyritic sulfur rejection. The last increment for the recovery of coarse or other hard-to-float particles is invariably accompanied by the flotation of much pyritic sulfur.

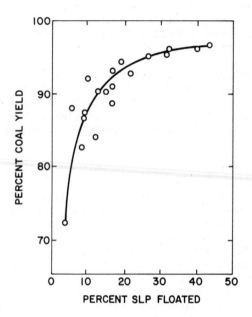

American Institute of Mining,
Metallurgical and Petroleum Engineers

*Figure 4.   Coal yield vs. liberated pyritic sul-
fur (SLP) recovery for Lower Kittanning coal
(4, 23)*

Acknowledgements

The author wishes to thank the American Iron and Steel Institute, The Pennsylvania Science and Engineering Foundation, and the Appalachian Regional Commission for financial support in conducting these studies. Acknowledgement is made to the following former and present graduate students in The Mineral Processing Section, The Pennsylvania State University, who have contributed data from which this report was drawn: C. M. Bonner, V. Choudhry, W. C. Hirt, C. J. Im, T. J. Olson and R. C. Rastogi.

Literature Cited

1. *Miner. Yearbk., 1973* (1975).
2. Mitchell, D. R., *Trans. AIME* (1948), *177*, 315.
3. Hoffman, L., Aresco, S. J., Holt, E. C., Deurbrouck, A. W., "Engineering/Economic Analysis of Coal Preparation with SO_2 Clean-up Process for Keeping Higher Sulfur Coals in the Energy Market," in *Second Symposium on Coal Preparation*, J. F. Boyer, Jr., Ed., Bituminous Coal Research Inc., Monroeville, Pa. (1976).
4. Aplan, F. F., "Coal Flotation," in *A. M. Gaudin Memorial International Flotation Symposium*, M. C. Fuerstenau, Ed., AIME, New York (1976).
5. Brown, D. J., "Coal Flotation," in *Froth Flotation - 50th Anniversary Volume*, D. W. Fuerstenau, Ed., AIME, New York (1962).
6. Zimmerman, R. E., "Flotation," in *Coal Preparation*, J. W. Leonard and D. R. Mitchell, Eds., 3rd ed., AIME, New York (1968).
7. Miller, F. G., Podgursky, J. M., Aikman, R. P., "Study of the Mechanism of Coal Flotation and its Role in a System for Processing Fine Coal," *Trans. AIME* (1967), *238*, 276.
8. Im, Chang J., unpublished data (1977).
9. Miller, F. G., "The Effect of Froth Sprinkling on Coal Flotation Efficiency," *Trans. AIME* (1969), *244*, 158.
10. Gaudin, A. M., *Flotation*, 2nd ed., McGraw-Hill, New York (1957).
11. Yancey, H. F., Taylor, J. A., "Froth Flotation of Coal; Sulfur and Ash Reduction," *U.S. Bur. Mines Rep. Invest.* (1935), *3263*.
12. Miller, F. G., "Reduction of Sulfur in Minus 28 Mesh Bituminous Coal," *Trans. AIME* (1964), *228*, 7.
13. Zimmerman, R. E., "Flotation of Bituminous Coal," *Trans. AIME* (1948), *117*, 338.
14. Miller, K. J., "Coal - Pyrite Flotation," *Trans. AIME* (1975), *258*, 30.
15. Hirt, W. C., Aplan, F. F., unpublished data (1977).
16. Rastogi, R. C., Aplan, F. F., "Coal Flotation as a Rate Process," accepted for publication in *Trans. AIME*.

17. Bushell, C. H. G., "Kinetics of Flotation," Trans. AIME
     (1962), 223, 266-278.
18. Arbiter, N., Harris, C. C., "Flotation Kinetics," in Froth
     Flotation - 50th Anniversary Volume, D. W. Fuerstenau, Ed.,
     AIME, New York (1962).
19. Rastogi, R. C., Hirt, W. C., Aplan, F. F., "An Evaluation of
     Several Variables Influencing the Rate of Coal Flotation,"
     submitted to Trans. AIME.
20. Rastogi, R. C., Hirt, W. C., Aplan, F. F., unpublished data
     (1977).
21. Bonner, C. M., Im, C. J., Choudhry, V., Aplan, F. F., "The
     Influence of Oil on Pyritic Sulfur Rejection During Coal
     Flotation," submitted to Trans. AIME.
22. Bonner, C. M., Aplan, F. F., "Frother Comparisons in the
     Flotation of Coal," accepted for publication in Trans. AIME.
23. Bonner, C. M., Hirt, W. C., Aplan, F. F., unpublished data
     (1977).
24. Aplan, F. F., "Mineral Processing - Evaluation to Indicate
     Processing Approach," Section 27.3 in SME Mining Engineering
     Handbook, A. B. Cummins and I. A. Given, Eds., pp. 27, 15-28,
     86-88, New York (1973).
25. Bonner, C. M., Hirt, W. C., Im, C. J., Aplan, F. F.,
     unpublished data (1977).

A Comparison of Coal Beneficiation Methods

SEONGWOO MIN[1] and T. D. WHEELOCK

Iowa State University, Department of Chemical Engineering and Nuclear Engineering, Energy and Mineral Resources Research Institute, Ames, IA 50011

Although iron pyrites and other minerals are removed from coal on an industrial scale almost exclusively by gravity separation methods at the present time, other beneficiation methods are coming into use. Among the developing methods, froth flotation (1,2,3) is the foremost, although the oil agglomeration method (4,5,6) is also promising. Both of these methods take advantage of the difference in surface properties of coal and inorganic mineral particles suspended in water to effect a separation. In the first method the hydrophobic coal particles are removed from the hydrophilic mineral particles by selective attachment to a mass of air bubbles, while in the second method the coal particles are selectively coated and agglomerated by fuel oil and then recovered by screening.

While gravity separation methods are well suited for removing coarse mineral particles from coal, they are generally ineffective for removing microscopic particles. On the other hand, both the froth flotation and oil agglomeration methods offer the potential for recovering and separating coal fines from microscopic impurities, thus complementing the gravity separation methods. However, none of these physical separation methods are effective unless the mineral impurities are first liberated or freed from the coal. Although mechanical crushing and/or grinding have always been used industrially to unlock impurities, the results have not always been satisfactory. Chemical comminution has been proposed as a means to unlock the impurities (7,8,9) and to solve this problem. This method of comminution uses specific chemical agents such as anhydrous ammonia to fragment coal. Since fragmentation occurs along bedding planes and boundaries between coal and mineral matter, the mineral impurities tend to be freed more completely for a given size reduction than would result from mechanical comminution (7).

[1]Present address: Battelle Columbus Laboratories, Columbus, Ohio 43201

In the work described here, high–sulfur bituminous coals from two Iowa strip mines were subjected to a series of 16 different treatments to compare the effectiveness of various beneficiation methods. These treatments involved different combinations of size reduction methods (crushing, pulverizing, grinding, and chemical comminution) and of physical separation methods (gravity separation, froth flotation, and oil agglomeration). The results are compared below on the basis of product yield and on the percentage reduction in sulfur and ash brought about by the treatments.

Experimental

The following methods, equipment, and materials were used for the experimental investigation:

Roll Crushing. Lump coal (4 cm x 0) was crushed to 6 mm top size by passing it through a bench–scale double roll crusher manufactured by Smith Engineering Works, Milwaukee, WI.

Pulverizing. Previously crushed coal was pulverized to -35 mesh by a Mikro-Samplmill manufactured by Pulverizing Machinery Division, American-Marietta Co., Summit, NJ.

Ball Milling. Previously pulverized coal was ground to -400 mesh size in a ceramic jar mill. For this operation 200 g coal, 1000 g water, and 1900 g flint pebbles were placed in a 5.7-L jar mill, and the mill was then run for 20 hr.

Chemical Comminution. For this operation 500 g lump coal (4 cm x 0) were placed in a 2000-ml Erlenmeyer flask which was then placed in a cold bath of dry ice and methanol and cooled to -70°C. Liquid anhydrous ammonia was then added to the flask until the coal was immersed in the liquid. After the coal had soaked for 1.0 hr, the flask was removed from the cold bath and was placed in a well ventilated hood where the ammonia evaporated. When the odor of ammonia could no longer be detected, the comminution step was completed.

Gravity Separation. To effect the gravity separation of coal and mineral matter, 500 g crushed coal (6 mm x 0) were added to 2000 ml tetrachloroethylene (specific gravity = 1.613) in a large beaker placed in a well ventilated hood. The mixture was stirred by hand to insure wetting of all particles, and then it was allowed to stand for 30 min. The float product was subsequently skimmed off and placed on a 100-mesh sieve to allow any adhering liquid to drain away. The float product was then placed in a drying oven at 100°C for 4 hr.

Froth Flotation. To conduct a froth flotation test, 200 g pulverized coal (-35 mesh) were added to 2000 ml tap water in a bowl of a laboratory model Wemco Fagergren flotation cell. With the agitator running, the pH of the slurry was lowered below 5 by adding 10 ml acid solution containing 10 vol % concentrated hydrochloric acid. Both kerosene (1.0 ml) and methyl isobutyl carbinol (0.5 ml) were added to the agitated slurry, and the air

flow to the cell was set at 9.3 L/min. The resulting froth was
collected until it appeared that no more coal was being recovered.
The material remaining in the bowl was poured out, the froth product was put back in the bowl, and fresh tap water was added to
bring the slurry volume to 2000 ml. With the agitator running,
the pH of the slurry was raised above 9 by adding 40 ml base
containing 5 wt % potassium hydroxide. The coal was then refloated without adding further reagents and using the same air flow
rate as before. A two-stage separation was made because a low pH
seemed to favor the removal of ash whereas a high pH seemd to
favor the removal of pyrite.

Oil Agglomeration. Oil agglomeration tests were carried out
with a 14-speed kitchen blender (Sears Insta-Blend Model 400)
which held up to 1200 ml fluid. For an agglomeration test, 200 ml
of an aqueous slurry containing 10 wt % coal was placed in the
blender together with 10 ml of solution containing 0.2 wt % sodium
carbonate. The sodium carbonate not only increased the pH of the
slurry but also served as a dispersing agent for the clay particles. The slurry was agitated for 5 min at the lowest speed. An
emulsion of fuel oil and water was then added to the coal slurry,
and the agitation continued for another 5 min at the same speed
to form agglomerates. The emulsion was prepared by combining
2.0 ml of a mixture of No. 1 fuel oil (86 vol %) and No. 5 fuel
oil (14 vol %), the mixture having a specific gravity of 0.83,
with 200 ml tap water and emulsifying the mixture with an ultra-
sonic vibrator. The agglomerated coal slurry was poured into a
1000-ml separatory funnel whereupon the coal floated to the surface, and the refuse particles settled to the bottom. The water
and refuse were drained out through the bottom opening, and the
agglomerated coal was put back in the blender and mixed with
200 ml fresh tap water. After agitating the mixture for 2 min at
the lowest speed, the coal slurry was poured back into the separatory funnel where the agglomerated coal was recovered again.
This washing operating was repeated once more to reduce entrapped
impurities.

Chemical Analysis Methods. Coal samples were analyzed for
sulfur and ash by the standard ASTM procedures (10).

Materials. Coal from two Iowa strip mines was used for this
investigation. A channel sample from the ICO mine and a run-of-
mine sample from the Jude mine were the source of the materials
used. The proximate analysis and sulfur distribution of each of
these samples are shown in Table I. Although the sulfur and ash
contents of these samples were widely different, the samples represented coal of the same rank (high volatile C). Investigation
of the coal microstructure with a scanning electron microscope
revealed substantial amounts of finely disseminated microcrystals
of iron pyrites (11,12).

Each coal sample was crushed to 4 cm top size and then was
divided into three size fractions (4 cm x 1 cm, 1 cm x 48 mesh,
and 48 mesh x 0). Each size fraction was then float-sink tested

Table I.   Composition of Coal from ICO and Jude Strip Mines

	Percent by Weight	
Type of Analysis	ICO	Jude
Proximate analysis		
volatile matter	44.0	39.4
fixed carbon	43.2	37.6
moisture	4.5	8.8
ash	8.3	14.2
total	100.0	100.0
Sulfur distribution		
pyritic	2.41	2.97
sulfate	0.05	0.44
organic	0.99	3.53
total	3.45	6.94

at various specific gravities using organic liquids of known specific gravity.   The standard Bureau of Mines procedure was used for this test (13).   The data for the different size fractions were combined to provide the composite washability analysis for 4 cm x 0 coal shown in Table II.

Treatment Results

The sequence of steps involved in each of the 16 treatments which were applied to each of the two coal samples is shown in Figure 1.   The first treatment was the simplest and involved crushing with the roll crusher, pulverizing with the Mikro-Samplmill, and oil agglomeration.   The second treatment included a ball milling step in addition to the other steps.   The third and fourth treatments included a froth flotation step.   In the fifth through eighth treatments the crushed coal was subjected to gravity separation before being pulverized and otherwise treated as in the first four treatments.   In the last eight treatments the coal was chemically comminuted before being conducted through the roll crusher.   Following the chemical comminution step, the pattern of treatments was the same as for the first eight treatments.   The final step of each treatment was an oil agglomeration step.

After each separation step within any given treatment, the weight of coal recovered was measured after drying the material

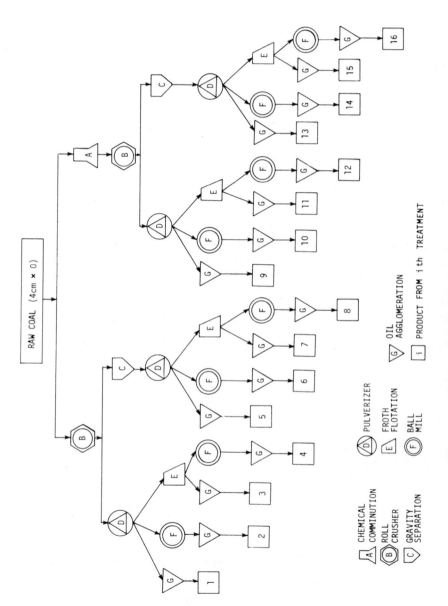

Figure 1.   Flow diagram of 16 different treatments

Table II. Composite Washability Analysis of Coal (4 cm x 0) from
           ICO and Jude Strip Mines

Product Fraction	Direct Data (%)			Cumulative Data (%)		
	Weight	Ash	Sulfur Total	Weight	Ash	Sulfur Total
ICO Coal						
Float 1.30	78.7	5.36	1.80	78.7	5.36	1.80
1.30 - 1.35	6.3	11.98	2.74	85.0	5.85	1.87
1.35 - 1.40	2.8	15.63	3.66	87.8	6.16	1.93
1.40 - 1.45	2.9	16.84	4.18	90.7	6.50	2.00
1.45 - 1.50	2.0	20.47	4.83	92.7	6.80	2.06
1.50 - 1.55	1.5	22.50	4.01	94.2	7.05	2.09
1.55 - 1.60	0.5	25.39	6.46	94.7	7.15	2.11
Sink 1.60	5.3	42.05	17.57	100.0	9.00	2.93
Jude Coal						
Float 1.30	46.5	5.67	5.07	46.5	5.67	5.07
1.30 - 1.35	19.0	12.03	4.83	65.5	7.51	5.00
1.35 - 1.40	7.6	15.84	5.24	73.1	8.38	5.03
1.40 - 1.45	7.1	19.94	6.41	80.2	9.40	5.15
1.45 - 1.50	3.9	23.80	7.91	84.1	10.07	5.28
1.50 - 1.55	3.0	28.30	7.53	87.1	10.70	5.36
1.55 - 1.60	1.7	31.51	6.79	88.8	11.10	5.38
Sink 1.60	11.2	48.92	12.84	100.0	15.33	6.22

overnight in an oven at 80°-100°C. A small sample of the dried
coal was subsequently analyzed for ash and pyritic sulfur. The
percentage reduction in either ash or sulfur content was found
for each separation step and for the overall treatment by using
the relation:

$$\text{Reduction (\%)} = \frac{\text{content of feed} - \text{content of product}}{\text{content of feed}} \times 100$$

The yield of coal for each separation step and the total yield
for the overall treatment were determined as follows:

$$\text{Yield (\%)} = \frac{\text{dry weight of product}}{\text{dry weight of feed}} \times 100$$

In the case of oil-agglomerated coal, the yield determined in this

manner included the small amount of oil which was not vaporized
during oven drying. Therefore, to express the yield on an oil-
free as well as on a moisture-free basis, the calculated yield
was reduced by 2%. It was assumed that losses of materials in the
crushing and pulverizing steps was negligible.

Figures 2, 3, and 4 show the cumulative effects of the different separation steps within each treatment as well as the overall results of each treatment. In general the results were quite
varied between treatments and between coals. Thus in the case of
ICO coal, the overall yield varied among the different treatments
between 74 and 96%, the overall reduction in pyritic sulfur content between 12 and 87%, and the overall reduction in ash content
between 22 and 72%. Similarly in the case of Jude coal, the overall yield varied among the different treatments between 75 and
92%, the overall reduction in pyritic sulfur content between 31
and 88%, and overall reduction in ash content between 34 and 84%.

Although the first separation step of a multistep treatment
produced the greatest reduction in level of impurities, subsequent separation steps also removed significant amounts of sulfur and ash (Figures 3 and 4). Hence, the separation methods
appeared to be complementary, particularly when used in conjunction with particle size reduction.

Among the various treatments, treatment 16, which included
all of the comminution and separation steps, produced the cleanest product from ICO coal. This product, recovered with an overall
yield of 78%, contained only 2.3% ash and 0.3% pyritic sulfur which
represented an overall reduction of 72% in ash content and 87%
in pyritic sulfur content. Treatments 8, 12, and 14 were nearly
as effective in removing sulfur and ash from ICO coal and provided higher yields than treatment 16.

Treatments 8, 14, and 16 were about equally effective in
removing sulfur and ash from Jude coal. The product of these
treatments contained about 0.4% pyritic sulfur and 2.2-2.7% ash
which represented an overall reduction of 86-88% in pyritic sulfur content and 81-84% in ash content. However, treatments 14
and 16 provided a larger overall yield (84%) than treatment 8
(75%).

The relative efficiency of the various treatments is illustrated by Figures 5 and 6 in which the overall yield of product is
plotted against the corresponding pyritic sulfur or ash content.
The upper curve in each diagram is drawn through the points representing the treatments which provided the highest yields for the
corresponding levels of impurities. Thus for ICO coal treatments
2, 10, 12, 14, and 16 were the most efficient from the standpoint
of sulfur removal and treatments 2, 6, 8, 14, and 16 from the
standpoint of ash removal. Similarly for Jude coal treatments 10,
11, 12, and 14 were the most efficient from the standpoint of sul-
fulr removal and treatments 10, 12, 14, and 16 from the standpoint
of ash removal. For the most part, these treatments had two
things in common. Thus, except for treatment 11 applied to Jude

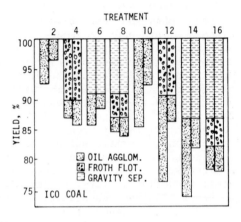

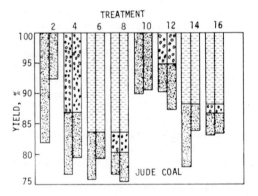

Figure 2. Weight yield of product from the
different treatments

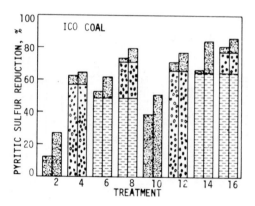

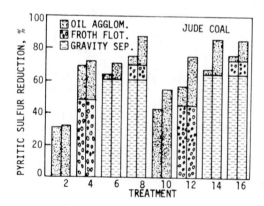

*Figure 3. Reduction in pyritic sulfur content
provided by the different treatments*

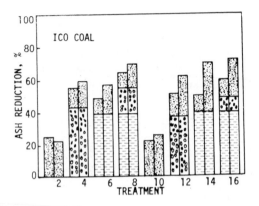

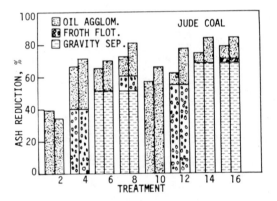

Figure 4. Reduction in ash content provided
by the different treatments

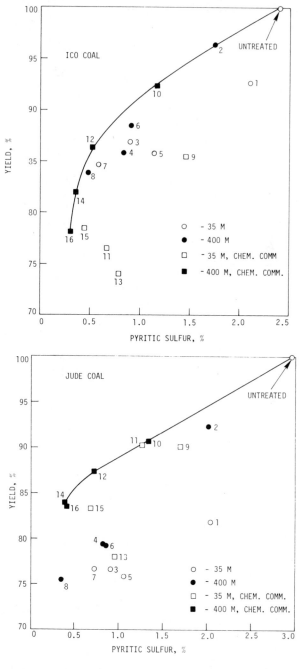

Figure 5. Overall yield vs. final pyritic sulfur content of the product from the different treatments

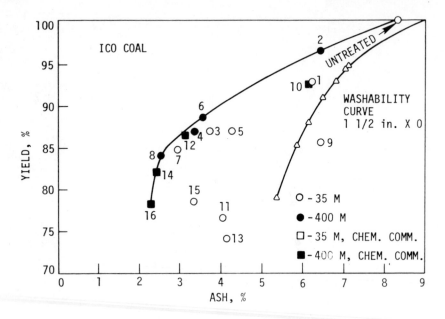

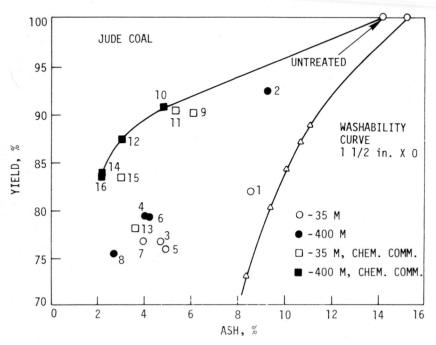

Figure 6. Overall yield vs. the ash content of the final product from the different treatments

coal, all of the treatments involved fine grinding before the oil
agglomeration step. In addition, except for treatments 2 and 6
applied to ICO coal, all of the treatments involved chemical
comminution.

The upper curve in each diagram of Figures 5 and 6 illustrates the trade off between overall yield and level of impurities
in the product. The yield fell off as the level of impurities
was reduced through application of treatments of greater and
greater complexity. With ICO coal the drop in yield was quite
precipitous when the pyritic sulfur content was reduced below
0.5% or the ash content below 2.5% (Figures 5 and 6). Therefore
for this coal, treatments 14 and 16, which provided the cleanest
coal but at considerable sacrifice in yield, would be especially
hard to justify.

Nearly all of the treatments described by Figure 1 were more
efficient than single-stage gravity separation of 4 cm x 0 coal.
The yield and corresponding ash content provided by gravity
separation of the coarse materials in liquids of different specific gravity are represented by the washability curves in Figure 6.
For either kind of coal the yield fell off sharply as the ash
content was reduced to lower levels by separation in lighter
liquids.

A comparison of the results of the first separation step of
each of the various treatments indicates the relative efficiency
of the different separation methods which were used. Thus in the
case of ICO coal, the froth flotation step of treatments 11 or 12
reduced the pyritic sulfur content more than the first separation
step of any other treatment, and the froth flotation step of
treatments 3 or 4 reduced the ash content more than the first
separation step of any other treatment. Also the respective
froth flotation steps provided the highest yields for the level
of impurities attained. The gravity separation step of treatments
5-8 was nearly as efficient in reducing the ash content of ICO
coal as the froth flotation step of treatments 3 or 4, but it was
not as efficient in reducing the sulfur content. However, gravity
separation was applied to coarser material than froth flotation.
In the case of Jude coal, the gravity separation step of treatments 13-16 provided the greatest reduction in pyritic sulfur and
ash contents of any of the first-step separations as well as the
highest yield for the level of impurities attained. The next
most effective first-step separation was that provided by oil
agglomeration of ball-milled Jude coal in treatment 10. This
step reduced the ash content of Jude coal almost as much as the
gravity separation step and also produced a higher yield. On the
other hand, it was not nearly as effective in reducing the
pyritic sulfur content as the gravity separation step of treatments 13-16.

It has already been noted that generally the most efficient
treatments involved fine grinding and chemical comminution. To
examine the effect of fine grinding further, the overall results

Table III.   Effect of Grinding on Overall Results

	Excluded (Odd Trt.)	Included (Even Trt.)
ICO coal		
ash reduction	47	55
pyritic sulfur reduction	58	67
yield	83	87
Jude coal		
ash reduction	65	71
pyritic sulfur reduction	61	71
yield	82	84

of the even-numbered treatments which included fine grinding and
the overall results of the odd-numbered treatments which excluded
this step were averaged separately. The two sets of values which
are presented in Table III indicate that on the average the treatments which included fine grinding produced a cleaner product in
greater yield than the treatments which did not include it.
Therefore it appears that fine grinding is a very effective means
for improving the separation of pyrite and other ash-forming
minerals from both ICO and Jude coals.

Similarly to study the effect of chemical comminution
further, the overall results of the last eight treatments which
included this step and the results of the first eight treatments
which excluded this step were averaged separately. The two sets
of values (Table IV) indicate that on the average the treatments
wich included chemical comminution provided a cleaner product
than the treatments which did not. Also with Jude coal, but not
with ICO coal, a larger yield was provided on the average by the
treatments which included this step. Therefore, at least for Jude
coal, chemical comminution is an effective method of improving
the separation of mineral impurities.

The effectiveness of the gravity separation step was also
evaluated further by averaging the overall results of the treatments which included it and the results of the treatments which
excluded it separately. The average reduction in both pyritic
sulfur and ash was much greater for those treatments which
included this step than for those which excluded it (Table V).
Although the average product yield was lower for the treatments
which included gravity separation, the penalty in yield was
rather modest for the large reduction in level of impurities
which was achieved.

Table IV.   Effect of Chemical Comminution on Overall Results

	Excluded (Trt. 1-8)	Included (Trt. 9-16)
ICO coal		
ash reduction	50	52
pyritic sulfur reduction	55	70
yield	88	82
Jude coal		
ash reduction	63	73
pyritic sulfur reduction	63	68
yield	80	86

Table V.   Effect of Gravity Separation on Overall Results

	Excluded (Trt. 1-4,9-12)	Included (Trt. 5-8,13-16)
ICO coal		
ash reduction	41%	62%
pyritic sulfur reduction	51%	74%
yield	88%	82%
Jude coal		
ash reduction	60%	76%
pyritic sulfur reduction	54%	77%
yield	86%	80%

Table VI. Effect of Froth Flotation on Overall Results

	Excluded (Trt. 1,2,5,6, 9,10,13,14)	Included (Trt. 3,4,7, 8,11,12,15,16)
ICO coal		
ash reduction	40%	62%
pyritic sulfur reduction	50%	75%
yield	87%	83%
Jude coal		
ash reduction	61%	74%
pyritic sulfur reduction	56%	75%
yield	84%	82%

The effectiveness of the froth flotation step was also evaluated by comparing the average overall results of the treatments which included this step with the average results of the other treatments Table VI). Here again the treatments which included this method of separation provided a much cleaner product than those which did not include it while experiencing only a modest reduction in yield. The effectiveness of froth flotation was very similar to that of gravity separation, although the two methods were applied to different sizes of coal.

The results of the individual steps of oil agglomeration and froth flotation were averaged to compare the effectiveness of one method against the other. Also in the case of oil agglomeration, the results of agglomerating −35 mesh coal were averaged separately from the results of agglomerating −400 mesh coal to determine the effect of particle size on this method of separation. From the data presented in Table VII it can be seen that the oil agglomeration step was much more effective when it was applied to −400 mesh coal than when it was applied to −35 mesh coal; not only did a cleaner product result, it was recovered in a larger yield. In addition the oil agglomeration step applied to −400 mesh Jude coal was more effective on the average than the froth flotation step applied to −35 mesh material. However, in the case of ICO coal the results were mixed with froth flotation appearing to have the edge with regard to sulfur removal but not with regard to ash removal.

Conclusions

The laboratory application of 16 different treatments in-

Table VII.   Oil Agglomeration Step vs. Froth Flotation Step

	Oil Aggl. -35 mesh (Trt. 1,5, 9,13)	Oil Aggl. -400 mesh (Trt. 2,6, 10,14)	Froth Flot. -35 mesh (Trt. 3,7, 11,15)
ICO coal			
ash reduction	20	34	30
pyritic sulfur reduction	16	40	51
yield	89	95	93
Jude coal			
ash reduction	36	47	31
pyritic sulfur reduction	23	44	35
yield	88	93	94

volving size reduction and physical separation to high-sulfur
coal containing substantial amounts of finely disseminated microcrystals of iron pyrites provided several interesting and important results. Comparison of these results with a standard
washability analysis showed that most of the treatments produced
a cleaner coal for a given yield than could be obtained by
gravity separation alone of 4 cm x 0 size coal. In this regard
the treatments which failed to produce coal with a lower sulfur
content were generally those which involved only size reduction
and oil agglomeration.

Treatments involving two and sometimes three methods of
separation in sequence proved particularly effective. Thus the
pyritic sulfur content of two Iowa coals was reduced 85-86% with
an overall yield of 82-84% by treatment 14 which included chemical
comminution, roll crushing, gravity separation, fine grinding,
and oil agglomeration. This sulfur reduction was considerably
higher than that provided by single-stage separation. Thus a
maximum reduction of 66% in the pyritic sulfur content of ICO
coal was realized during the froth flotation step of treatments
11-12 and of 64% in the pyritic sulfur content of Jude coal during
the gravity separation step of treatments 13-16.

Each of the separation methods used in these treatments
proved effective in itself. Moreover the methods seemed to com-
plement each other, particularly when used in conjunction with
particle size reduction.

Chemical comminution generally improved the separation
efficiency of the various treatments and fine grinding the
separation efficiency of the oil agglomeration method of separa-

tion in particular.

Acknowledgement

This report is based on a paper which was presented at the NCA/BCR Coal Conference and Expo III, Louisville, KY, Oct. 19-21, 1976. The work was sponsored by the Iowa Coal Project and conducted in the Energy and Mineral Resources Research Institute at Iowa State University.

Literature Cited

1. Aplan, F. F., this volume, in press.
2. Cavallaro, J. A., Deurbrouck, A. W., this volume, in press.
3. Zimmerman, R. E., Min. Cong. J. (1964) 50 (5), 26.
4. Capes, C. E., Smith, A. E., Puddington, I. E., CIM Bull. (1974) 67, 115.
5. Capes, C. E., McIlhinney, A. E., Coleman, R. D., Trans. Soc. Min. Eng., AIME (1970) 247, 233.
6. Brisse, A. H., McMorris, W. L., Jr., Trans. Soc. Min. Eng., AIME (1958) 211, 258.
7. Howard, P. H., Datta, R. S., this volume, in press.
8. Datta, R. S., Howard, P. H., Hanchett, A., NCA/BCR Coal Conference and Expo III (Oct. 19-21, 1976), Louisville, KY.
9. Howard P. H., Hanchett, A., Aldrich, R. G., Symposium II Clean Fuels from Coal, (June 23-27, 1975), Institute of Gas Technology, Chicago, IL.
10. Book ASTM Stan., (1976) Part 26, "Gaseous Fuels; Coal and Coke; Atmospheric Analysis," Methods D2015-66, D2492-68, D3174-73, D3177-75.
11. Greer, R. T., this volume, in press.
12. Greer, R. T., "Nature and Distribution of Pyrite in Iowa Coal," Joint Meeting of the Electron Microscopy Society of America and Microbeam Analysis Society, (August 9-13, 1976), Miami, FL.
13. Methods of Analyzing and Testing Coal and Coke, U.S. Bur. Min. Bull. (1967) 638.

Dry Table—Pyrite Removal from Coal

D. C. WILSON

FMC Corp., Santa Clara, CA 95052

Particle segregation within a flowing granular bed during passage through material handling equipment or during transfer from one piece of equipment to another is commonplace. The most readily observable of these is size segregation. Usually such segregations are undesirable in a system because of the potential of unstable processing and an interruption of product continuity.

Three factors determine the extent of particle segregation in a moving particle bed: the physical configuration of the material handling equipment, the forces which convey the particles through that equipment, and the differences between the particles in one or more of their physical properties (size, shape, bulk density, resiliency and surface roughness). The equipment described below, specifically designed as a separator for dry particulates, combines these factors to exploit the inherent segregation within moving particle beds. The equipment can be used to remove pyrite and other ash-forming minerals from coal; experimental results of cleaning bituminous and subbituminous coals with it are reported.

Equipment Description

Figure 1 contains a perspective drawing of the dry table, being developed by FMC Corporation, and a cross section through the unit illustrating the particle bed. The drive unit for the table shown is an electromechanical exciter of the type used for vibrating feeders. In fact, this recently developed coal cleaning unit is a feeder, but with the following design differences:

(1) The deck surface is short but very wide.

(2) The coal is inserted at one side of the feeder's deck.

(3) The conveying force is reversed; it feeds the material into the table's backwall.

(4) The particle bed's net flow is from one side of the feeder to the other side.

(5) The deck is nonsymmetrical about the vertical plane passing through its center of gravity and the exciter's line of

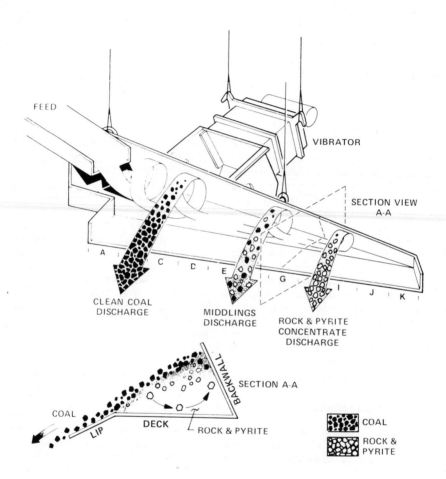

FEED

VIBRATOR

SECTION VIEW
A-A

A   C   D E   G   I J   K

CLEAN COAL
DISCHARGE

MIDDLINGS
DISCHARGE

ROCK & PYRITE
CONCENTRATE
DISCHARGE

COAL

BACKWALL

SECTION A-A

LIP   DECK   ROCK & PYRITE

COAL

ROCK &
PYRITE

Figure 1.   Dry table principle; schematic view

drive.

Coal is fed onto the longest side of the unit, and the conveying force from the drive moves the particles towards the backwall. A large pile of particles forms against the backwall, filling the entire trough. Gravity moves the particles on the pile's surface down the open slope as the conveying force con‐ tinues to drive the underlying material against the backwall. The result is the continuous overturning of the bed. The pressure of the incoming feed forces the overturning bed to flow across the deck away from the feed side in a helical motion, and because the deck's length diminishes (tapers) in this direction, the toe of the pile is being continuously discharged.

Simultaneously, size and bulk density separations are occur‐ ring in the overturning bed. The large or low–density particles move into a spiralling path that migrates toward the toe of the pile (Section A–A, Figure 1) whereas the small or high–density particles move into a smaller spiral and concentrate towards the backwall. Those particles that are both large and of low density (coal) advance past the large and high–density particles (rock and pyrite) and prevail in obtaining positions at the toe of the pile. Also small particles of pyrite will concentrate at the backwall in preference to small particles of coal.

The overall resulting discharge from the horizontal deck por‐ tion of the unit is a series of staggered particle size gradations of different densities (when bulk densities are directly related to apparent densities). To avoid this overlapping of the size gradations of the rock and pyrite with the coal, the feed to the unit is presized to definite size ranges. For the coal, rock, and pyrite separation, the generalized top size to bottom size of the feed particles in any one pass is a 4-to-1 ratio (8 x 2in., 2 x 1/2 in., etc.). The optimum ratio, however, is a function of bulk density, shape, resiliency, and surface roughness differences between the coal and rock-pyrite for a particular coal seam and mining conditions.

The particles discharge from the nearly horizontal deck onto an attached downward sloping surface referred to as the "discharge lip". This lip can make further separations based on particle shape, resiliency, and surface roughness if desired. The shape separation is based on the cubical coal particles being unstable on the discharge lip and the near tabular rock and pyrite parti‐ cles being stable when the unit is vibrating. The unstable coal will thus be discharged by rolling off the lip while the tabular rock and pyrite are conveyed back up the lip into the pile. The surface roughness of the highly mineralized particles is greater than that for the clean coal particles. This additional roughness helps to convey the rock and pyrite back into the deep particle bed whereas the slick coal tends to slip off the lip. Generally, the resiliency of the coal is greater than that of the rock particles. The conveying vibrations causes the more resilient coal particles to bounce and assure their unstability on the discharge lip.

The above describes the major separations occurring at specified locations on the dry table. Even though particle shape, resiliency, and surface roughness differences are not included in the description of the overturning particle pile, they do contribute to minor separations within the pile. Likewise, particle bulk density and size differences are involved in the separation on the lip, but these are of minor importance when compared to the differences in particle shape, resiliency and surface-roughness. With the separation driving forces of the five known physical property differences operating simultaneously, it is highly unlikely that all five are working together to arrive at the desired coal cleaning. Therefore, most coal cleaning separations are a compromise in which the dry table's operating parameters are adjusted to enhance the desired physical property separations and depress the undesirable separations.

The following physical property differences list describes how the dry table parameter controls influence the degree of the individual separations.

Size. For coal cleaning, the feed to the dry table is pre-screened into size ranges to minimize the naturally strong size segregation effects in a moving particle bed. The narrower the size range, the less size segregation effect on the overall separation. Systems can be devised for post-screening but at a sacrifice of the dry table's throughput rate.

Bulk Density. For coal cleaning a moderately deep pile is best, as described below. If the frequency and/or amplitude is increased above a normal level, then the particle bed becomes quite porous and all particles will exhibit a lighter bulk density, thus reducing the effective bulk density differences.

Shape. Forward or backward tilting of the dry table causes the discharge lip to be at different angles relative to the horizontal which determines what shapes are rejected or retained on the table. A discharge lip angle of $-12°$ is generally used for the separation of cubic coal from tabular rock-pyrite. If the rock-pyrite is present as round particles, then a less steep discharge lip is used.

Resiliency. By increasing the unit's frequency and/or stroke above the normal level, the more resilient particles in the bed will exhibit a lower bulk density in the pile section and simultaneously become more unstable on the lip section.

Surface Roughness. The type of surface finish on the discharge lip combined with the particle's own surface roughness will determine if it is discharged or retained on the table. A very hard slick surface will discharge nearly all particle types while a very hard rough surface will retain nearly all particle types.

The throughput rate and bed retension time are controlled by altering the cross-slope (side to side of unit). Increasing the cross-slope, by lowering the feed side of the unit with respect to its opposite side, will cause lower throughput rates and large bed retension times.

There is a close relationship between the dry table's vertical stroke component and frequency, with some limitations dictated by the particle sizes being separated and the optimization of throughput rates. Of obvious importance is the horizontal conveyance of particles to form the overturning pile. The depth of that pile is then, to a large extent, determined by the vertical conveyance. And the pile depth is important because very deep piles favor size segregation while piles of moderate depth favor bulk density segregations. For coal cleaning, therefore, a fixed drive angle of 30° was selected as the best combination of horizontal and vertical conveyances in forming the desired piles of moderate depth. The frequency is related to the stroke; with increasing stroke length, the frequency must be reduced to maintain the desired tabling activity at the selected drive angle:

$$\text{Stroke (Frequency)}^2 = \text{Constant}$$

The stroke should not be larger than the smallest particle diameter to be separated or particle mixing becomes a factor in the separation. However, using the largest size stroke possible is desirable because the throughput rate increases with increasing stroke:

$$\text{Stroke} \propto \text{(Throughput Rate)}^2$$

Since most of the bed particles are in contact with each other rather than with the dry table surfaces, the character of the particle discharge is a function of the bed composition itself as well as the above described parameters. If the rock-pyrite content of the feed increases, a proportional increase in rock-pyrite would be expected throughout all the discharge products, including the cleanest coal product. This phenomenon has been demonstrated many times for a wide composition of feed which indicates that the dry table's separation principle is based on a probability function.

The dry table has a discharge similar in character to that of a wet concentration table in that it is a gradation from a clean coal product through mineralized particles to pyrite along the discharge edge. This dry method of separation is functional over a broad span of particle sizes. The limiting factor for the minimum size particle is the formation of particle agglomerations because of electrostatic charges or surface moisture. No limiting factor has been encountered for the maximum size particles. The present practical range in coal preparation is 1/8 - 8 in., the

lower limit being dictated by screening practices and the upper
limit by the size of the present 15 ft prototype dry table.

A small   12 in. lab unit can predict the performance of the
larger 8- and 15-ft units on a particular raw coal with a feed of
proportionally sized particles because of the probability base of
the dry table's separation principle. Therefore, small-scale coal
cleaning tests are made first with the lab unit to determine the
parameter settings and to reduce field test time requirements.
Standard washability data of raw coal is of little value in pre-
dicting what the performance of the dry table would be because
this data is based on apparent densities alone.

Most of the dry table experience is in the reduction of the
ash content of coals. However, there has been a recent increase
of interest in the use of the dry table as a method for sulfur
reduction.

Experimental

Runs were made with the following-described samples by passing them through one of the dry table units in one pass and
collecting the discharge as multiple products. In Figure 1, an
11-product discharge is shown, A through K, where the discharges
are of equal increments spaced along the discharge lip. Each
discharge product was analyzed for ash, pyritic sulfur, and BTU
content, following the accepted ASTM methods D-271 and D-2492.
The two coal samples used are described in Table I.

Table I. Sample Descriptions and Test Parameters

	Bituminous	Subbituminous
Source		
location	Raton, NM	Fruitland, NM
condition	wet processed (heavy media)	raw (direct from pit)
mine	underground	surface
Analysis		
BTU/lb	13,460	8,060
Ash (%)	10.1	25.8
Pyritic sulfer (%)	0.44	0.19
Sample run		
particle size (range)	1/4 in. x 8 mesh	2 x 1/2 in.
surface moisture	~0%, as received	~0%, as received
Dry table parameters		
location	Santa Clara, CA (FMC)	Navajo Mine
unit size	12-in (lab unit)	8-ft (pilot only)
throughput rate	0.01 tons/hr.	3 tons/hr.
drive type	electromagnetic	electromechanical
frequency (Hz)	60	18
stroke (in.)	0.03	0.25
drive angle	30°	30°

Results

The analytical results of the runs with the bituminous and
subbituminous samples were used to construct the distribution
curves shown in Figure 2. The horizontal axis for both the upper
and lower portion of the graph represents the discharge from the
dry table as the 11-discharge products. In the upper portion, the
vertical axis gives the recovery as a percentage of the original
feed for heating content (BTU), ash, and pyrite. The clean coal
product is the accumulation of discharge products starting at the
far left (percentage on left vertical axis), and the reject starts
on the far right (right vertical axis). The data points plotted
on the graph are for runs which had an eight-product discharge.
In the lower portion of the graph, separate curves are plotted for
product and reject which show the distribution of the pyritic
sulfur as lb per million BTUs.

Discussion

As with all coal cleaning equipment, the performance is a
function of the coal being cleaned. The dry table is no exception
to this and can be considered more sensitive because it uses as
many as five of the particles' physical property differences for
separation rather than just the density difference alone. Also
affecting the separation is the degree to which the major constituents of the raw coal (clean coal, rock, and pyrite) are
liberated from each other. Therefore, a gradation of property
differences for all the particles' properties, as well as for
size, can exist in the granular feed to be segregated. Then to
this add the dry table's probability base for particle segregation, and the complexity of in depth interpretation of the dry
table's results can be invisioned.

The pyrite in the two examples selected for this discussion
is unliberated and of relatively low concentration. The lettered
discharge products A through D are divisions of the dry table's
discharge arbitrarily selected for analytical and discussion
purposes, and the large number of product divisions or their
specific boundaries, need not be used in actual coal cleaning
applications.

Subbituminous Coal. The only preparation this coal received
prior to being fed to the dry table was presizing into 4:1
size ranges. The sequence of the distribution curves, (Figure 2,
top) show that in this three component system, clean coal (BTU) -
rock (ash) - pyrite, the major separation is between the clean
coal and the rock. The pyrite-ash separation is reversed to what
would be expected for the more dense pyrite, which further demonstrates that the pyrite is not liberated. There are three zonal
types of discharge from the dry table with this coal. In the
first zone, product discharges A and B, the coal contains low ash
and has the pyrite mainly associated with the coal. In the second

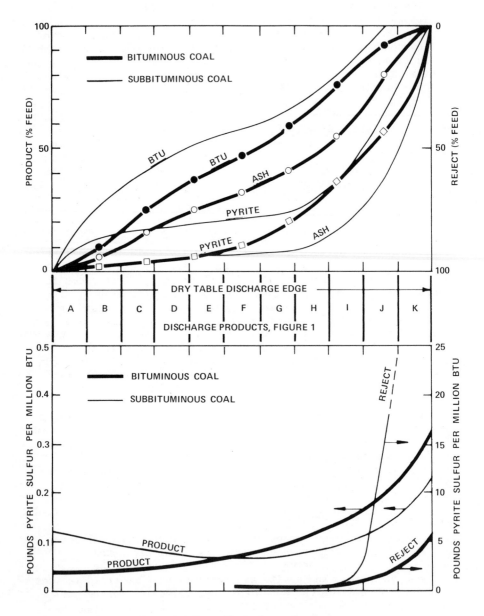

Figure 2.   Dry table discharge distribution

zone, product discharges C through G, the coal contains low pyrite and ash, but the pyrite is associated with the ash. In the third zone, product discharges H through K, the discharge contains a coal and rock mixture where the pyrite is associated with both coal and rock at higher concentrations.

Selecting discharges A through I as a clean coal product and J and K as reject gives a 90% recovery of the coal's potential heating content and removals of 75% of the ash and 50% of the pyrite. The compositions of the product and reject are given in Table II.

Table II.   Selected Subbituminous Coal Product

	Product	Reject
Yield (%)	72	28
BTU/lb	10,360	2,160
Ash (%)	10.8	64.1
Pyritic sulfur (%)	0.13	0.32

With the discharge split into just a clean coal product and reject, the well known compromise must be made between recovering as much clean coal as possible while rejecting most of the rock and pyrite. With the dry table, however, it is possible to have as many products as are reasonable. Therefore, one could select the zone of low ash content A through G as the clean coal product and H through J for retreatment, where there is a mixture of both rock and coal, and K as the reject which essentially contains no usuable heat (Table III).

Table III.   Selected Subbituminous Coal Product and Recycle

	Product	Retreatment	Reject
Yield (%)	47	37	16
BTU/lb	11,260	7,390	0
Ash (%)	4.7	30.0	78
Pyritic sulfer (%)	0.08	0.27	0.29

Both the clean coal product and the reject are desirable in this arrangement, and its success depends upon the character of the retreatment discharge. In this particular case, the retreatment is a mixture of clean coal and liberated rock with a small amount of middlings (this would be expected because the dry table's particle segregation has a probability base). The retreatment can be recycled through the same unit or sent to another unit for a second pass.

Bituminous Coal. This coal is the product from a preparation plant, and therefore there is essentially no liberated pyrite or rock present. The sequence of the curves shows that the unliberated pyrite is unevenly distributed among the coal particles and that the major separation is between the clean coal and pyrite. The ash and BTU curves are very similar, except for a slight

difference in slopes, showing that there is a near-constant
inherent ash in the coal for the discharge products A through H.
The pyrite curve is quite different in shape and shows small
amounts of pyrite in discharge products A through D, increasing
amounts in E to J, and substantial quantities in K. There are
four zonal types of discharge for this particular separation. In
the first zone, discharge products A through D, the coal has a
minimum ash and pyrite content. In the second zone, discharge
products D to H, the coal contains a minimum of ash but has an
increasing pyrite content. In the third zone, discharge products
I through J, the ash and pyrite content progressively increase in
the coal. In the fourth zone, discharge product K, the coal is
highest in both ash and pyrite and contains all the misplaced sink
material from the wet washing process.

The specific gravity difference between the first and second
zones is quite small, so the separation is most likely caused by
the other physical properties of the particles. Since the ash
difference is also small, it is assumed that the presence of the
pyrite is related to the physical property differences which the
dry table can distinguish for separation purposes. The suggested
method of processing is to collect the discharge from discharge
products A to H as a clean coal product and I to K for retreatment
(Table IV).

Table IV.   Selected Bituminous Coal Product and Recycle

	Product	Retreatment
Yield (%)	75	25
BTU/lb	13,750	12,600
Ash (%)	7.5	17.5
Pyritic sulfur (%)	0.22	1.08

For this coal sample the retreatment discharge should not be processed as a second pass on the dry table because little benefit
would be realized in ash and pyrite reduction. The best approach
would be to recycle this material to the wet preparation plant
after crushing.

Conclusion

The dry table can reduce the sulfur content of a coal through
pyrite removal. The extent of the coal-pyrite separation will be
a function of pyrite liberation and the physical property differences between the free flowing coal and pyrite particles. However, there are cases where even coal containing unliberated pyrite
can be separated into coal products of low and high pyritic sulfur
contents.

The dry table is best used as a rougher, since it has a separation similar to that of a Baum Jig. It can be used alone or in
conjunction with existing coal cleaning equipment. And it is

especially applicable where the use of water is restricted because of limited supply, freezing, or costly treatment prior to discharge or reuse.

9

Magnetic Desulfurization of Some Illinois Basin Coals

HAYDN H. MURRAY

Dept. of Geology, Indiana Univ., Bloomington, IN 47401

High extraction magnetic filtration (HEMF) is used success-
fully to process kaolin (1). This is the first successful commercial application of a new level of magnetic separation equipment and processing technology which resulted from the joining
of four major concepts (2).

1. Discovery of the importance of retention time in mineral
separation.

2. Development of very high gradient matrix collectors.

3. High intensity fields in wet magnetic separator (up to
20 kG).

4. Modern design of large high field magnets.

Use of longer retention time permits finely divided particles to migrate and be captured by a magnetized collection surface. The canister in the magnet is filled with a matrix of
steel wool, screens made of sharp thin ribbons, or other filamentary material which provides very high gradients. Modern
electronic and magnet technology led to the design of a magnet
with a high field throughout a large cavity. A diagrammatic
sketch of a large high intensity magnet is shown on Figure 1.
The diameter of the canister can be up to 84 in. with a height
of 20 in. Up to 120 tons of kaolin per hour can be processed
through the 84-in. unit. Fabrication of equipment larger than
84 in. is feasible, but the problems involved in shipping and
for on site fabrication are such that it is probably more efficient to consider multiple installations of 84-in. machines.
High extraction magnetic filtration is very successful in
removing iron and titanium impurities from kaolin. Potential
applications for its use for beneficiation of other industrial
minerals and coal have been demonstrated by Murray (3,4,5).
Present HEMF equipment uses electromagnets to generate fields

of 20 kG. Power consumption of this equipment is in the range
of 400–500 kW.

The present HEMF equipment is optimized for separation of
slurry containing fines below 200 mesh and preferably below 20
microns. Other matrix types can be substituted for stainless
steel wool to accomodate coarser feed materials (up to 20 mesh)
including Frantz screens, loosely packed coarse steel wool, steel
shot, steel filings, and other filamentary material. New developments are underway in matrix design and composition which can
greatly enhance HEMF technology.

Magnetic Desulfurization of Coal

The earliest work concerning the reduction of sulfur in coal
by magnetic separation was described in a German patent by
Siddiqui in 1957 (6). Yurovsky and Remesnikov (7) published a
paper in 1958 reporting that coal pulverized finer than 16-mesh
size subjected to a thermal steam-air treatment made the pyrite
more magnetic, which enhanced beneficiation when processed in
a specially built magnetic separator. Sulfur reductions of 85,
74.9, and 70% were reported. Perry (8) reported that fine pyrite
(65 – 70 mesh) treated in steam-air atmosphere at temperatures
of 570° – 750°F for varying times up to 10 min. increased the
quantity of pyrite amenable to magnetic separation with increasing intensity of treatment. Kester (9, 10,) demonstrated that
sulfur could be reduced to a greater extent by making a high-
intensity magnetic separation directly on raw untreated coal
without using the thermal pretreatment step. Thus, by pulverizing the coal to a typical power plant size and by magnetically
separating the coarse 48 by 200-mesh size fraction, significant
sulfur reduction was achieved.

Kester reported that pyritic sulfur accounts for 40 to as
much as 80% of the sulfur content of most coals (9). Gluskoter
and Simon (11) reported that the mean total sulfur content in
474 analyses was 3.57% in coals from Illinois, and the mean value
of pyritic sulfur in these same coals was 2.06%. They found that
there is on an average approximately 150% more pyritic sulfur in a
sample as there is organic sulfur.

Macroscopic pyrite occurs in coal in: veins, usually thick
and filmlike along vertical joints; in lenses that are extremely
variable in shape and size; in nodules or balls; and in disseminated crystals and irregular aggregates. Microscopic pyrite
occurs as small globules and blebs, fine veinlets, dendrites,
small euhedral crystals, cell fillings, and replacement plant
material.

Kester, Leonard, and Wilson (12) reported that the mass
susceptibility of powdered pyrite was 4.53×10^6 cgs units.
Another value commonly used for the magnetic susceptibility of
pyrite is 25×10^{-6} electromagnetic units per cm. The strength
of magnetism, which can be induced into a mineral depends upon

the permeability of the mineral according to the equation:

$$\underline{B} = u\underline{H}$$

Where--$\underline{B}$ = magnetic induction in gauss in the mineral, $\underline{u}$ = permeability of the mineral, and $\underline{H}$ = magnetic field intensity in gauss. Therefore the susceptibility is:

$$\underline{B}/\underline{H} = 1 + 4 \; \pi\underline{K}$$

Where--$\underline{K}$ = magnetic susceptibility expressed in electromagnetic units $\overline{cm}$/gm/sec. If the value of $\underline{K}$ is positive, the mineral is termed paramagnetic and experiences a force which tends to attract it in the direction of increasing magnetic gradient. If $\underline{K}$ is negative, the mineral is diamagnetic and experiences a repulsive force. Ferromagnetic minerals, such as iron, experience strong magnetic forces in the direction of increasing magnetic gradient and thus have very large positive values of $\underline{K}$. Coal is diamagnetic (13), and pyrite is paramagnetic. Thus, if the coal is crushed and pulverized fine enough to liberate the pyrite, a good magnetic separation is possible.

A recent study by Kindig and Turner (14) reported on a new process for removing pyritic sulfur and ash from coal. The pulverized coal is treated with iron carbonyl vapor which puts a thin skin of magnetic material on the pyrite and ash but does not affect the coal. Thus magnetic separators yield a non-magnetic coal low in sulfur and ash and a magnetic fraction high in sulfur and ash.

Experimental

The coal samples used for this report were pulverized so that 90% passed through a 200-mesh sieve. The samples were slurried at 30% solids for the wet magnetic tests. Frantz screens made from thin sharp ribbons of 430 magnetic stainless steel were used as the matrix in the canister. For the wet magnetic tests retention times of 30, 60, and 120 sec. were used for one series, and multiple passes with a retention time of 30 sec. each were used for a second series. For the dry tests the series were run using gravity feed with multiple passes.

The magnetic separator used in this study was a pilot plant model manufactured by Pacific Electric Motor Co. of Oakland, California. The HEMF equipment includes a filter canister which can be packed with a filamentary magnetic material. The matrix material can easily be changed depending on the particle size of the feed slurry. The canister is surrounded by hollow conductor copper coils which energize the entire canister volume to a field up to 20 kG. The copper coils are surrounded by a box-like enclosure of steel plate which completes the magnetic circuit. The unite operates at power levels of 200 kW.

The coal samples were pulverized in a laboratory hammer mill and a small ring mill. After the pulverized coal was slurried with the aid of a dispersant (sodium hexametaphosphate), it was agitated with a Lightnin mixer and pumped up through the canister with a peristaltic pump at controlled rates. Approximately 1 lb of coal was run through the canister for each test. Samples of -100 mesh were run and compared with the -200 mesh samples and much better results were obtained with the finer samples.

The coals used for this report were commercially mined coals from the Illinois Basin. These are Coals V and VI from Illinois and Indiana. Table I shows the sulfur content of the various samples.

Table I. Sulfur Content (%)

Coal	Total Sulfur	Inorganic Sulfur	Organic Sulfur	% Ash
Indiana V (Warrick Co.)	4.63	2.44	2.19	12.8
Indiana VI (Warrick Co.)	4.17	2.20	1.97	11.4
Illinois V (Wabash Co.)	3.59	2.39	1.20	10.3
Illinois VI (Williamson Co.)	1.98	1.02	0.96	7.1

Figures 2 and 3 indicate the sulfur reduction obtained with increasing retention time and up to three passes through the magnet using wet separation methods. Figure 4 shows the sulfur reduction obtained using a dry separation technique. The data show that the best results as far as sulfur reduction is concerned were attained using a slurry and three passes through the magnet, each with a retention time of 30 sec. Table II is a summary of the sulfur reduction obtained using both wet and dry separation methods with -200 mesh samples.

Table II. Sulfur Reduction (%)

Coal	Total S	Total S in Product	Inorganic S in Product	% Inorganic S Removed	% Ash
Indiana V	4.63	3.00[a]	0.81	67	6.4
Indiana V	4.63	3.30[b]	1.11	55	7.9
Indiana V	4.63	3.78[c]	1.59	25	9.2
Indiana VI	4.17	2.30[a]	0.10	85	6.2
Indiana VI	4.17	2.45[b]	0.25	78	7.1
Indiana VI	4.17	3.31[c]	1.01	39	8.8
Illinois V	3.59	1.96[a]	0.83	65	5.8
Illinois V	3.59	2.18[b]	0.99	59	6.3
Illinois V	3.59	2.87[c]	1.67	30	8.6
Illinois VI	1.98	1.15[a]	0.21	79	4.1
Illinois VI	1.98	1.29[b]	0.32	69	4.9
Illinois VI	1.98	1.57[c]	0.61	40	5.6

[a] Wet-three passes;   [b] 120-sec retention;   [c] dry-three passes.

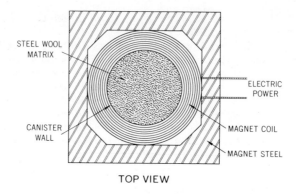

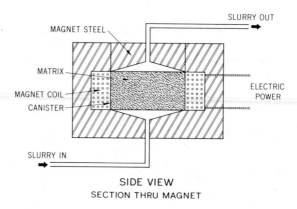

Figure 1. *Diagrammatic side and top view of HEMF unit*

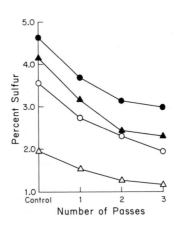

Figure 2. *Sulfur content after 1, 2, and 3 passes at 30 sec retention each (wet)*

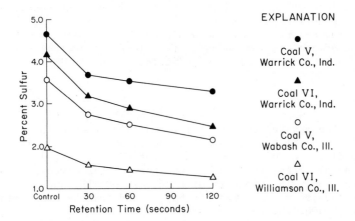

Figure 3.   *Sulfur content wtih increasing retention time (wet)*

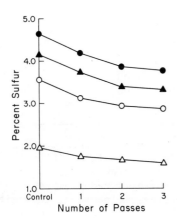

Figure 4.   *Sulfur content after 1, 2, and 3 passes (dry)*

One sample of Coal V from Indiana was pulverized so that 90% of its particles passed 325 mesh and using 3 passes with 30-sec retention each, 93% of the pyritic sulfur was removed. Further tests on fine grinding and optimization of the test conditions are now being carried out in the author's laboratories. In addition to the sulfur reduction, ash reduction is being measured. The loss of coal in the magnetic fractions varied from 6 to 14% and is related to the size and distribution of the pyrite in the coal.

Economics

Quinlan and Venkatesan (15) recently discussed the economics of coal preparation coal cleaning processes comparing jig versus heavy media plant circuits. The operating cost of the jig plant was $0.85 per clean ton and for the heavy media circuit $1.25 per clean ton. The capacity of each was 500 TPH, and the capital cost of the jig circuit was $6,000,000 and for the heavy media circuit $8,500,000.

To design a cleaning circuit to product 500 TPH of coal would require five 84-in. magnets. The capital cost (installed) would be approximately $8,000,000.

Table III. Operating Costs of HEMF For 500 Tons/hr. Capacity

	Cost per Hour	Cost per Ton
Amortization of installed separators over 10 years, 80,000 hr.	$100.00	$0.20
Magnet power (2000 kW @2¢ kW·h)	40.00	0.08
Pumping and flushing power (1000 kW @2¢ kW·h)	20.00	0.04
Labor	15.00	0.03
Maintenance	10.00	0.02
Total	185.00	0.37

The cost per ton of magnetic cleaning is relatively low compared with the other two methods cited. In addition to the potential removal of 70 - 90% of the inorganic sulfur, the ash content of the coal would be substantially reduced. A high percentage of the following impurities, if present, in addition to pyrite, would be removed magnetically because all of these minerals and rocks have a mass susceptibility higher than pyrite except for limestone: siderite, limonite, ferrous and ferric sulfate, limestone, clay and shale, and sand.

Much additional research and development work must be done to substantiate the preliminary results reported in this paper. Several studies are underway in the author's laboratories at

Indiana University. Since coal is to become the major energy
source in the U.S. in the foreseeable future, magnetic cleaning
of coal looks as if it will be a viable method of processing
which can provide a low-sulfur, low-ash coal.

Conclusions

High extraction magnetic filtration (HEMF) is a proven commercial process used extensively in the processing of kaolin.
Fine pulverization to -200 mesh or preferably to -325 mesh to
liberate the pyrite from the coal is necessary to achieve optimum
separation. From 65 - 90% of the inorganic sulfur can be separated from a coal slurry by HEMF processing. The estimated
cost per ton is low and compares favorably with other methods of
pyrite removal. The resultant coal product from the HEMF process
is relatively clean of inorganic sulfur and ash content.

Abstract

High extraction magnetic filtration (HEMF) has potential
application to remove a high percentage of inorganic pyritic
sulfur from coal. Newly developed high intensity separators with
high intensity (20 kG) and high gradient collectors can remove
up to 90% of the pyritic sulfur under optimum conditions. A
canister filled with a filamentary matrix of sharp stainless
steel screens provides high gradients and pyrite in pulverized
coal (-200 mesh) will be attracted and held. Coal is diamagnetic,
and pyrite is paramagnetic, and if the pyrite can be liberated
from the coal, it can be magnetically filtered out in commercial
practice. Coals V and VI from Indiana and Illinois were used
for this study. These coals have a total sulfur content of
between 1.98 and 4.63% of which approximately 60% is pyritic
sulfur. Both dry and wet magnetic separation methods were used.
Wet methods were more successful in that up to 90% of the pyrite
was removed whereas the best results from the dry separation was
40% removal. The cost per ton of a 500 ton/hr installation would
be of the order of $0.50 per ton and would produce a relatively
low-sulfur, low-ash product.

Literature Cited

1.   Iannicelli, J., "High Extraction Magnetic Filtration of
     Kaolin Clay," Clays Clay Miner., (1976) 24, 64-68.
2.   Iannicelli, J., "Assessment of High Extraction Magnetic
     Filtration," Special Report-RANN-National Science Foundation
     (1974).
3.   Murray, H.H., "Beneficiation of Selected Industrial Minerals
     and Coal by High Intensity Magnetic Separation," IEEE Trans.
     Magn., (1976) MAG-12, 498-502.

4. Murray, H.H., "High Intensity Magnetic Beneficiation of Industrial Minerals-A Survey," Preprint 76-H-93 Amer. Inst. Min. Engrs. (1976).

5. Murray, H.H., "High Intensity Magnetic Cleaning of Bituminous Coal," Proc. National Coal Research Symposium - Louisville, Ky., pp 74-81, Oct. 1976.

6. Siddiqui, S., "Desulfurization and Concentration of Coal," German Patent 1,005,012, 1957.

7. Yurovsky, A.Z., Remesnikov, I.D., "Thermomagnetic Method of Concentrating and Desulfurizing Coal," Coke Chem. (1958) 12, 8-13.

8. Perry, H., "Potential for Reduction of Sulfur in Coal by Other Than Conventional Cleaning Methods," Symposium-Economics of Air Pollution Control, AICHE 59th Nat. Meeting 1966.

9. Kester, W.M., Jr., "Magnetic Demineralization of Pulverized Coal," Mining Eng. (1965) 17.

10. Kester, W.M., Jr., "The Effect of High Intensity Magnetic Cleaning on Pulverized High Sulfur, Bituminous Coals," Thesis, School of Mines, West Virginia Univ., 1966.

11. Gluskoter, H.J., Simon, J.A., "Sulfur in Illinois Coals," Ill. State Geol. Surv., Circ. (1968) 432.

12. Kester, W.M., Jr., Leonard, J.W., Wilson, E.B., "Reduction of Sulfur from Steam Coal by Magnetic Methods," Tech. Report 31, Coal Research Bureau, West Virginia Univ. (1967).

13. Wooster, W.A., Wooster, N., "The Magnetic Properties of Coal," Bull. British Coal Utilization Research Assoc. (1944).

14. Kindig, J.K., Turner, R.L., "Dry Chemical Process to Magnetize Pyrite and Ash for Removal from Coal," AIME Preprint 76-F-306 (1976).

15. Quinlan, R.M., Venkatesan, S., "Economics of Coal Preparation" AIME Preprint 76-F-108 (1976).

This study was supported by a grant from the Electric Power Research Institute. The chemical analyses were done by Nily Sofer in the chemistry lab, Geology Department at Indiana University.

Desulfurization of Coals by High-Intensity High-Gradient Magnetic Separation: Conceptual Process Design and Cost Estimation

C. J. LIN and Y. A. LIU

Magnetochemical Engineering Laboratory, Department of Chemical Engineering, Auburn University, Auburn, AL 36830

It is well known that the main difficulty in increasing the utilization of coal in the United States lies in the pollution problem, and emission levels of sulfur oxides and ash particles from coal burning facilities must be reduced to meet stringent environmental standards. Although the particulate emission standards can generally be met by using electrostatic precipitators, there apparently exists no accepted technology for controlling sulfur oxide emissions in flue gases (1). Thus, there has been a growing effort recently to develop effective and economical alternatives to flue gas desulfurization. One of the most attractive alternatives is the precombustion cleaning of coal. Several new physical and chemical methods for removing sulfur and ash from coal prior to its combustion have already been proposed and are currently under intensive further development (2). An important physical method for cleaning coal that appears to hold much promise is the well established magnetic separation technique. Previous experimental investigations have indicated clearly that most of the mineral impurities in coal which contribute to the pyritic sulfur, the sulfate sulfur, and the ash content are paramagnetic. These sulfur-bearing and ash-forming minerals, if sufficiently liberated as discrete particles, can be separated from the pulverized diamagnetic coal by magnetic means (3-8). Indeed, the technical feasibility of the magnetic cleaning of coal has been demonstrated in a number of previous studies, with substantial amounts of sulfur and ash removal reported (9,10).

During the past few years, the magnetic cleaning of coal has been given new impetus with the introduction of a new level of magnetic separation technology, the high-intensity high-gradient magnetic separation (HGMS). The HGMS technology was developed around 1969 for the wet separation of feebly magnetic contaminants from kaolin clay (11-17). A typical HGMS unit for this wet application is shown schematically as Figure 1a. The electromagnet structure consists of energizing coils and a surrounding iron enclosure. The coils in turn enclose a cylindrical, highly magnetizable working volume packed with fine strands of strongly

ferromagnetic packing material such as ferritic stainless steel
wool. With this design, a strong field intensity of 20 kOe
can be generated and distributed uniformly throughout the working
volume. Furthermore, by placing in the uniform magnetic
field the ferromagnetic packing materials which increase and dis-
tort the field in their vicinity, large field gradients of the
order of 1-10 kOe/μm can be produced. In the wet beneficiation of
kaolin clay, the HGMS unit is used in a batch or cyclically operated process like a filter. The kaolin feed containing the feebly
magnetic contaminants in low concentrations is pumped through the
stainless steel wool packing or matrix of the separator from the
bottom while the magnet is on. The magnetic materials (mags)
are captured and retained inside the separator matrix, and the
nonmagnetic components (tails) pass through the separator matrix
and are collected as the beneficiated products from the top of
the magnet. After some time of operation, the separator matrix
is filled to its loading capacity. The feed is then stopped, and
the separator matrix is rinsed with water. Finally, the magnet
is turned off, and the mags retained inside the separator matrix
are backwashed with water and collected. The whole procedure is
repeated in a cyclic fashion. In general, if this batch process
is used in other wet applications where the magnetic materials
occupy a large fraction of the feed stream, the down time for
backwashing will be considerable, possibly necessitating the use
of one or more back-up separators. To overcome this problem which
is inherent in batch operations, a continuous process using a
moving matrix HGMS unit, called the Carousel separator, has been
proposed (12, 15-17) as shown schematically in Figure 1b. A number of pilot-scale studies of the wet beneficiation of kaolin clay
and iron ores using the Carousel separator have been reported
(17).

    Because of the very low costs and the outstanding technical
performance of HGMS demonstrated in the kaolin application, HGMS
was recently adapted in a bench-scale exploratory study (8) to
remove sulfur and ash from a finely pulverized Brazilian coal
suspended in water. Since then, other investigators have utilized pilot-scale HGMS units to desulfurize and deash water
slurries of some Eastern U.S. coals. For instance, results of
pilot-scale studies that demonstrated the technical feasibility
of the magnetic separation of sulfur and ash from water slurries
of pulverized Illinois No. 6, Indiana No. 5 and No. 6, Kentucky No. 9/14, Pennsylvania Upper Freeport and Lower Kittanning
coals have been reported (4-7, 18, 19). In particular, the
quantitative effects of major separation variables such as field
intensity, residence time, etc., on the grade, recovery and concentration breakthrough for the magnetic separation of sulfur
and ash from a water slurry of pulverized Illinois No. 6 coal
have been established experimentally, and can be predicted satis-
factorily by a newly developed mathematical model for HGMS (4,5,
7,13,18). Depending upon the types of coal and the separation

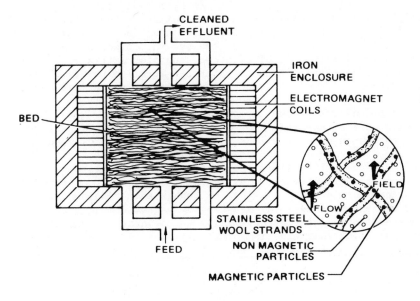

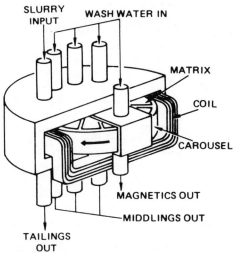

Figure 1. (a) (above) Cyclic high-gradient magnetic separator; (b) (below) carousel high-gradient magnetic separator

conditions used, the existing bench- and pilot-scale results have
shown that the use of single-pass HGMS was effective in reducing
the total sulfur by 40-55%, the ash by 35-45%, and the pyritic
sulfur by 80-90%, while achieving a maximum recovery of about
95% (10,18). These available results have indicated also that
both the grade and recovery of the separation can be enhanced
generally with the use of a larger separator matrix or by the
recycle of the tail products. Further detailed review of the
reported results on the magnetic cleaning of pulverized coals
in water slurries can be found in the literature (9,10,18). These
published data and other recent analyses (2,10,18,20-24) have
indicated that a significant portion of the U.S. coal reserves,
low enough in organic sulfur, can be magnetically cleaned for use
as an environmentally acceptable, low sulfur fuel. It has been
estimated that 100 million short tons of U.S. coals per year may
be magnetically cleaned. This amounts to over 17% of the total
U.S. production per year (10). Although the existing data have
not yet established total deashing by magnetic means, there is
some indication that by optimizing the separation conditions,
and enhancing the magnetism of ash-forming minerals, etc., the
effectiveness of magnetic separation of ash from coal can be improved (10).

Recent studies (4,5,9,10,25-27) have also suggested that
coal cleaning by HGMS could serve as a significant adjunct to
coal liquefaction processes. In particular, the technical feasibility of adapting HGMS as an effective alternative to conventional precoat filtration in the solvent refined coal (SRC) process has already been demonstrated by a bench-scale, exploratory study done at Hydrocarbon Research, Inc. (HRI). HGMS effec-
tively removed up to 90% of the inorganic sulfur from the liquefied SRC filter feed slurry of Illinois No. 6 coal, and about half
of the experimental runs conducted by HRI indicated over 87% inorganic sulfur removal (6,15,26). In general, the work done by
HRI showed that HGMS was less effective in ash removal but did
remove 25-35% of the ash. Quite recently, a pilot-scale HGMS
system to remove mineral residue from liquefied coal has been
designed and constructed by the authors (5,25). Typical results
from experiments conducted with the liquefied SRC filter feed
slurry of Kentucky No. 9/14 and Illinois Monterey coals have
been quite encouraging, indicating that HGMS could reduce the
total sulfur, ash and pyritic sulfur contents by as much as 70,
76, and 95%, respectively. Available data from the above bench-
and pilot-scale investigation have showed also that an even
greater deashing of the liquefied SRC filter feed can be achieved by improved separation conditions. A detailed discussion of
these results along with their technical implications can be
found in the literature (10,25,27). Furthermore, close examination of the inherent physical and chemical characteristics of
the hydrogenated products prior to the filtration step in the
SRC and other related liquefaction processes indicates that HGMS

may be developed as a practical mineral residue separation method. Hydrogenation will generally reduce a major portion of the pyritic sulfur to the highly magnetic pyrrhotite; and the sulfur-bearing and ash-forming minerals tend to be liberated more easily from the dissolved organic components in the filter feed slurry than from pulverized coal suspended in water. Also, the typical mean particle size of the filter feed sample is often less than 5 microns, which dictates the use of separation methods like HGMS capable of handling micron-size materials. All of these factors seem to suggest a very significant potential for utilizing HGMS to remove mineral residues from liquefied coal. For certain types of coal, even without further improvement in the magnetic removal of ash, magnetically cleaned SRC would be acceptable for use as a feed to boilers which already have electrostatic precipitators (28). This follows because the cost of solid-liquid separation in coal liquefaction is generally substantial, and moderately low-ash SRC should be less expensive (27-29). Indeed, a preliminary cost estimate for the magnetic desulfurization of liquefied coal based on the laboratory data obtained by HRI seems to support this observation (27).

The preceding discussion has indicated that the scientific and technical feasibility of magnetic desulfurization of both coal/water slurry and liquefied coal has been well established. Recently, several estimates of the cost of magnetic desulfurization have been reported in the literature (8,14,19,22,27). Because of the simplifying assumptions involved, as well as the technical performance specified and the estimation methods used, the results seem to be somewhat approximate in nature. In this paper, the latest data from pilot-scale studies of sulfur and ash removal from both liquefied coal and coal/water slurry by HGMS are used to design conceptual processes for magnetic desulfurization of coal. Estimates of magnetic desulfurization characteristics and conceptual process requirements as well as installation and processing costs are presented. In particular, the extent to which the processing conditions can affect magnetic desulfurization costs is examined. Finally, the results are compared with other approaches for the desulfurization of coal.

Magnetic Desulfurization of Coal/Water Slurry: Process and Costs

A conceptual process for the magnetic desulfurization of pulverized coal suspended in water by HGMS is shown schematically in Figure 2. A coal slurry of a fixed concentration is prepared first by mixing known amounts of pulverized coal, water, and a dispersant (wetting agent) like Alconox. The HGMS unit used here is the largest commercial unit now in use for producing high-quality paper coating clays. It is operated at a fixed field intensity of 20 kOe generated in an open volume 7 ft in diameter and 20 in. long. A stainless steel wool separator matrix having

94% voids is placed in the open volume. The coal slurry is
pumped through the energized separator matrix at a fixed residence time (flow velocity) until the matrix reaches its loading
capacity. After rinsing with water, the mags are sent to a
settling pond or a clarifier to recover water for re-use. The
tails are collected, dewatered, and dried. The specifications
of process streams for a typical case (Case B in Table I) are
also included in Figure 2, and the detailed operating conditions
for such a case are given in the Appendix.

By removing 80-90% of the pyritic sulfur magnetically and
achieving 85-90% recovery as was demonstrated by the results of
reported studies of magnetic desulfurization of pulverized
coals in water slurries (4-10,13,18,19), the process can be used
to clean about one-fifth of the recoverable U.S. coals with a
low organic sulfur content of 0.7-0.9 wt % to produce an environmentally acceptable fuel. A detailed documentation of the reserves and production of U.S. coals which may be magnetically
cleanable to 1 wt% total sulfur according to the seam, district
and county in each state, along with the total and organic sulfur
contents can be found in the literature (21).

Here, a reasonable range of add-on costs (excluding those
for grinding, dewatering and drying and refuse disposal) is estimated for the wet magnetic cleaning of coal slurries which have
desulfurization characteristics similar to those reported in the
recent studies (4-10,13,18,19). The method used to estimate the
costs of magnetic desulfurization is based on the technique used
by the Federal Power Commission Synthetic Gas-Coal Force to
estimate the cost of synthesis gas (29, 34). The investor capi-
talization method used in this approach is the discount cash
flow (DCF) financing method with assumed DCF rates of return such
as 15% after taxes. This method essentially determines the
annual revenue during the plant life which will generate a DCF
equal to the total capital investment for the plant. Several
major assumptions are included in the method (29,34):
(1) The plant life is assumed to be 20 years with no cash value
    at the end of life.
(2) A straight-line method is used to calculate the annual depre-
    ciation.
(3) Operating costs and working capital requirements are assumed
    to be constant during the construction period, and 100% equity
    capital is assumed.
(4) Total plant investment, return on investment during the plant
    life, and working capital are treated as capital costs in
    year zero (the year ending with the completion of start-up
    operations).
(5) Start-up costs are treated as an expense in year zero.
(6) 48% federal income tax is assumed.
Based on these assumptions, equations for calculating the unit
costs ($ per ton of coal processed) are suggested by the published
documents (29,34). They are summarized in the Appendix, in which

the detailed operating conditions and estimated costs for a typical example shown in Figure 2 are also given. The costs of major
installed equipment and the unit costs listed in the Appendix
are based on values as of June 1976. For instance, the costs of
pump and tank used were estimated first according to Ref. 35
and then brought up to date by multiplying by the CE plant cost
index ratio of (205/113.6), while the cost of the installed
HGMS unit with a separator matrix 7 ft in diameter and 20 in.
long was estimated to be $1.936 million (26).

The estimated capital investment and unit costs for four
typical cases, designated as A-D, are summarized in Table I.
Slurry velocities of 2.61 and 4.0 cm/sec, slurry concentrations
of 15,25 and 35 wt%, as well as separation duty cycles of 59.0-
77.9% are considered. These separation conditions illustrate
clearly the effects of slurry velocity and concentration as well
as separation duty cycle. For instance, comparison of cases A
to C shows that at the same slurry velocity and similar magnetic desulfurization characteristics the higher the slurry concentration, the cheaper will be the investment and unit costs.
While this observation is to be expected, there have been pilot-
scale tests which indicate that increasing the slurry concentration of pulverized Illinois No. 6 coal from 2.57 to 28.4 wt%
did not appreciably change the grade and recovery of the separation. Further effects of processing conditions, and other cost
factors on the unit costs are illustrated in Table II. It is
seen from the table that by doubling the amount of coal processed
per cycle relative to a fixed amount of stainless steel wool
packed in the separator matrix, the unit cost can be reduced by
about 15%. This result shows the importance of the separator
matrix loading characteristics on the costs of magnetic desulfurization. Another factor which affects the unit costs considerably is the washing time required in a complete separation
cycle. This is seen by comparing items 4 and 6 in Table II. In
particular, the computed results indicate that doubling the amount
of wash water required only leads to negligible increase (0.27-
0.60%) in unit costs. However, if both the amount of wash water
and the washing time are doubled, the unit costs are increased
by about 15%. The above observations clearly suggest an important economic incentive for further pilot-scale investigation
of the separator matrix loading and washing characteristics in
the magnetic desulfurization of coal/water slurry. Finally, item
7 of Table II shows that labor cost seems to be a significant
fraction of the unit cost. Fortunately, it is not expected that
the labor requirement will be doubled in actual commercial practice from the nominal case in Table I. This follows because
existing experience in the commercial cleaning of kaolin clays
by HGMS indicates that the labor requirements for both operation
and maintenance are minimal (11,22).

In Table III, the reported costs of magnetic desulfurization
expressed in terms of the capital and unit costs ($ per ton of

Table I.   Cost of Desulfurization of Coal/Water Slurry
           by HGMS Using Separator Matrix of 7-Ft
           Diameter and 20-In Length

		Case A	Case B	Case C	Case D
1.	Slurry velocity (cm/sec)	2.61	2.61	2.61	4.0
2.	Slurry concentration (wt %)	15	25	35	25
3.	Coal feed rate (ton/hr)	44.77	66.13	83.07	89.61
4.	Cycle time (min)	9.00	6.10	4.85	4.50
5.	Duty cycle (%)	77.9	67.4	59.0	59.6
6.	Tons of coal processed per cycle	6.7	6.7	6.7	6.7
7.	Unit costs ($ per ton coal processed)				
	U	2.083	2.369	1.109	1.067
	U_o	1.802	1.063	0.858	0.829
	U_{15}	3.676	2.412	1.967	1.880
8.	Capital investment per ton coal processed per year ($)	6.93	4.69	3.73	3.53

Table II.   Sensitivity Analysis of Unit Costs ($ Per Ton Coal Pro-
            cessed) of Desulfurization of Coal/Water Slurry by HGMS

	U_o		U_{15}	
	$	% Change	$	% Change
1. Basis: 2.61 cm/sec, 25 wt % slurry, and other conditions in the Appendix and Table I	1.0628	0.00	2.4117	0.00
2. Amount of coal processed per cycle doubled	0.9004	-15.28	2.0341	-15.66
3. 25% reduction in capital investment	0.9389	-11.66	1.9506	-19.12
4. Amount of wash water required doubled (washing time unchanged)	1.0691	+0.60	2.4181	+0.27
5. Cost of water increased 5/3 times (5¢/1000 gal)	1.0835	+1.95	2.4324	+0.86
6. Both amounts of wash water and washing time doubled	1.2256	+15.32	2.7883	+15.62
7. Labor requirement doubled	1.3587	+27.82	2.7077	+19.12

coal processed) are compared with the results of this study. The
costs given by Murray (19) were based on the existing cost estimate for kaolin beneficiation by HGMS given in Ref. 11. For a
residence time of 0.5 min, the coal feed rate to a commercial
HGMS unit with a separator matrix 7 ft in diameter and 20 in.
long was set at 100 tons per hr by Murray. This rate appears
to be higher than that expected in commercial practice. In
addition, the costs of labor and maintenance per HGMS unit
were estimated by Murray to be $1 and $2 per hr, respectively.
These costs also appear to be lower than those reported in
Ref. 11. Consequently, the costs estimated by Murray shown in
Table III especially the unit cost U_O ($0.37 per ton processed),
are believed to be lower than the actual costs. Next, while the
costs estimated by Oder (22) seem to be relatively comparable
to those obtained in this study, it is difficult to identify
clearly the differences between the two estimates. This follows
because specific details regarding the costs of major installed
equipment, cycle time, and washing time, etc., were not reported
in Ref. 22. Finally, the costs estimated by Trindade (8) are
also believed to be lower than the actual costs. In the cost
estimate by Trindade, a Carousel separator was used as a basis
for the estimation, although there have not yet been any test
data reported on the magnetic desulfurization of coal/water
slurry using a Carousel separator. Only the separator cost was
included in the capital cost in the analysis by Trindade, and it
was about one-half of the cost of installing an equivalent cyclic
HGMS unit. This led to the relatively low capital investment
per ton of coal processed ($0.82-$1.64) estimated by Trindade
as shown in Table III. The cost estimation method used by Trindade generally leads to lower unit costs. For instance, by
using Trindade's method, the unit cost U_O obtained in this work
for a slurry velocity of 2.61 cm/sec shown in Table III will be
reduced from $1.06 to $0.85 per ton of coal processed.

A comparison of approximate estimated capital and unit costs
of different pyritic sulfur removal processes currently under
active development (2,30,31,33) is given in Table IV. With
the exception of the MAGNEX process (31), all the methods
listed in Table IV are wet processes and require grinding of feed
coal, thus requiring relatively comparable grinding, dewatering,
and drying costs. This table indicates that the costs of wet
magnetic desulfurization by HGMS are attractive when compared
with those of other approaches, even after adding the necessary
costs of grinding, dewatering and drying. However, the above
comparison is only an approximate one, because of the difference
in the methods used to estimate the costs and in the desulfurization characteristics reported, etc. Based on the available cost
information for these pyritic sulfur removal processes (3,30,31,
33), it is not yet possible to carry out a rigorous comparison.

Magnetic Desulfurization of Liquefied Coal: Process and Costs

Table III. Comparison of Estimated Costs of Desulfurization of Coal/Water Slurry by HGMS

	Murray(19)	Oder(22)	Trindade(8)		This Work	
1. Slurry velocity (cm/sec)		1.7-3.4	2	4	2.61	4.0
2. Slurry concentration (wt %)			30		25	
3. Unit costs, $ per ton coal processed						
U_o	0.37	0.70-0.25	0.39-0.84	0.22-0.45	1.06	0.83
U_{15}		2.47-0.93			2.41	1.88
4. Capital investment per ton coal processed per year ($)	2.02	5.95-2.28	1.64	0.82	4.69	3.53

Table IV. Comparison of Estimated Approximate Capital and Unit Costs of Different Pyrite Removal Processes

Process	U_o [a]	U_{15} [a]	Capital Investment [b]
1. MAGNEX-Hazen Research, Inc.(31)	5.83	7.05	4.17
2. Froth flotation-Bureau of Mines(31)	2.77	4.47	5.71
3. Meyers-TRW Systems and Energy	6.0-14.0		13.80 (leaching only)
4. Ledgemont oxygen leaching-Kennecott Copper Corp.(30)	comparable to Meyers		11.30 (leaching only)
5. HGMS-This work, see Table I	0.83-1.06	1.88-2.41	3.53-4.69

[a] Unit costs expressed in $ per ton coal processed.
[b] Capital investment expressed in $ per ton coal processed per year.

A flow diagram of the conceptual process to remove the mineral residue from liquefied SRC by HGMS is shown in Figure 3. The HGMS unit used here is the same commercial separator used to desulfurize the coal/water slurry. The magnetic desulfurization of liquefied SRC is to be conducted at elevated temperature to reduce the viscosity of the coal slurry. Furthermore, the packed stainless steel wool matrix is also to be heated to the desired separation temperature during operation. The elevated temperature of the matrix will prevent the coal slurry from congealing and plugging the matrix. It is also necessary to insulate the heated portion of the matrix from the magnet windings. The insulated matrix is further surrounded by a water jacket. These provisions for heating, insulating, and cooling the separator matrix slightly reduce the actual working volume of the separator matrix and its diameter from 7.0 to 6.83 ft. In actual separation runs, the unfiltered liquefied SRC is pumped through the energized separator at a constant flow rate until the separator matrix reaches its loading limit. After rinsing with a process generated solvent, the matrix is backwashed with the same solvent with the magnet de-energized. The mags are sent to a hydroclone separator. The overflow from the hydroclone is recycled back to the wash solvent tank for re-use; while the underflow is sent to an evaporator to recover the solvent, and the residual solids are packed for other uses. The tails from the separator are sent to a vacuum column to recover the solvent for process recycle and the vacuum bottom is sent to a product cooler to produce the solidified SRC.

The conceptual process is designed to achieve the same extent of inorganic sulfur removal and matrix loading observed by HRI for slurry velocities of 0.25–14.0 cm/sec [26]. The specific magnetic desulfurization characteristics corresponding to these slurry velocities are summarized as the first three rows of Table V. According to a survey of 455 U.S. coal samples conducted by the Bureau of Mines, the average total and inorganic sulfur contents of these samples were 3.02 and 1.91 wt%, respectively [20]. Thus, if the hydrogenation step in the SRC and other related liquefaction processes can remove 70% of the organic sulfur, a reduction in the inorganic sulfur content by about 67% after hydrogenation will be sufficient to produce SRC with an emission level smaller than 1.20 lb SO_2/million Btu, assuming that the SRC has a heating value of 16,000 Btu/lb.

The same method of cost estimation summarized in the Appendix except for replacing the dispersant by steam with a nominal cost of $2/1000 lb was used to estimate capital investment and unit costs of the conceptual process. The results are presented in Table V. Here, the cost of major installed equipment such as the HGMS unit, wash solvent tank, feed surge tank, feed pump, flush pump and evaporator, etc., has been included. In Table VI, the effect of steam price on the unit cost U_o of magnetic desulfurization of liquefied coal is illustrated. Doubling the

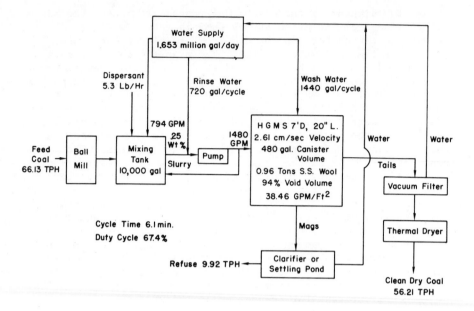

Figure 2.   *Desulfurization of coal/water slurry by HGMS*

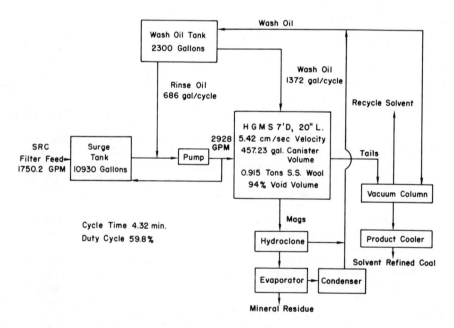

Figure 3.   *Desulfurization of liquefied coal by HGMS*

Table V.   Costs of Desulfurization of SRC Feed Coal by HGMS Using Separator Matrix of 6.83-Ft Diameter and 20-In Length

1.	Slurry velocity (cm/sec)	0.25	1.60	2.71	2.71	5.42	14.0
2.	Apparent % pyritic sulfur removal	90	87	78	74	67.7	66.4
3.	Cycle time (min)	45.86	9.15	5.84	11.33	4.32	2.39
4.	Duty cycle (%)	85.8	74.9	66.3	82.6	59.8	33.5
5.	Tons of SRC feed coal processed per cycle	22.52	24.71	23.63	57.21	31.48	25.25
6.	Filtration rate based on actual filtering time (GPM/ft^2)	3.74	23.56	39.91	39.91	79.82	206.2
7.	Unit costs, $ per ton coal processed						
	U_0	6.58	1.45	1.10	0.76	0.67	0.61
	$\overline{U}_{15}$	15.65	3.31	2.53	1.80	1.54	1.37
	$\overline{U}$	8.64	1.87	1.42	1.00	0.87	0.79
8.	Capital investment per ton coal processed per year	30.5	6.26	4.79	3.49	2.89	2.53

Table VI.   Effect of Steam Price on the Unit Cost for Desulfurization of SRC Feed Coal by HGMS

Slurry Velocity (cm/sec)	Unit Cost[a]		% Increase from
	Case A	Case B	Case A to Case B
0.254	6.580	6.600	0.3
2.71[b]	1.100	1.311	19.2
2.71[c]	0.763	0.850	11.4
5.42	0.670	0.836	24.8
14.00	0.614	0.811	32.1

[a] Unit cost expressed in $ per ton SRC feed coal processed
Case A: steam price = 2 $/1000 lb; Case B: steam price = 4 $/1000 lb.
[b] Amount of liquefied coal processed per cycle = 23.62 tons = 25.83 times weight of stainless steel wool.
[c] Amount of liquefied coal processed per cycle = 57.21 tons = 62.53 times weight of stainless steel wool.

Table VII. Approximate Comparison of Capital and Unit Costs
            of Different Solid-Liquid Separation Methods (29)

Method	$\underline{U}_o{}^a$	$\underline{U}_{15}{}^a$	Capital Investment[b]
1. Rotary drum filtration (SRC)	2.77	8.10	17.89
2. Pressure leaf filtration (SRC)	7.03	9.87	9.52
3. Two-stage centrifugation[c] (H-Coal)	2.57	7.57	15.38
4. Solvent precipation (H-Coal)	1.82	3.98	6.70
5. HGMS[c]			
5.4 cm/sec	0.68	1.54	2.89
2.7 cm/sec	1.10	2.53	4.79

[a] Unit costs expressed in $ per ton SRC feed coal processed.
[b] Capital investment expressed in $ per ton SRC feed coal processed per year.
[c] The ash content of separated product may not satisfy EPA specification, and the use of electrostatic precipitators may be needed.

steam price will increase the unit cost U_o by 0.3-32% in the range of slurry velocity considered. As steam is mainly used in the evaporator to recover the wash solvent, this comparison also implies that the higher the process throughput, the more expensive will be the operating cost for solvent recovery. Finally, a comparison of the approximate capital investment and unit costs of several solid-liquid separation methods, including precoat filtration, centrifugation, solvent precipitation and HGMS, is given in Table VII (29). This table shows that although the precoat filtration and solvent precipation can generally meet the stringent environmental standards for both sulfur and ash, the costs of these methods are higher than those for HGMS. Thus, there seems to be some economic incentive for using magnetically cleaned SRC as a feed to boilers which already have electrostatic precipitators. Obviously, additional development work is needed to support this observation firmly.

Acknowledgement

The financial support provided by the Auburn University Board of Trustees 1975-1976 Research Award, the National Science Foundation (Grant Nos. GI-38701 and AER 76-09300), the Gulf Oil Foundation, and the Engineering Experiment Station of Auburn University are gratefully acknowledged.

Abstract

The latest experimental data from pilot-scale studies of sulfur and ash removal from both pulverized coal suspended in a water slurry and from liquefied solvent refined coal by high gradient magnetic separation are used to design conceptual processes for desulfurization. Estimates of magnetic desulfurization characteristics and conceptual process requirements as well as installation and processing costs are presented. The results indicate that the magnetic processes appears to be attractive in terms of costs and performance when compared with other approaches to desulfurization.

Appendix

Basis for Estimating the Unit Costs of Magnetic Desulfurization of Coal/Water Slurry. The detailed operating conditions and estimated unit costs for a typical conceptual process for the magnetic desulfurization of coal/water slurry (case B in Table I) are illustrated as follows.

Operating Conditions

(1) Concentration of coal/water slurry = 25 wt%
(2) Superficial flow velocity = 2.61 cm/sec

(3) Stainless steel wool packing density = 6 wt%
(4) Amount of coal processed per cycle = 7 times weight of
    stainless steel wool
(5) Amount of rinse water required per cycle = 1.5 times volume
    of separator matrix
(6) Amount of wash water required per cycle = 3 times volume
    of separator matrix
(7) Flow velocity of rinse water = flow velocity of coal slurry
(8) Washing time per cycle = 1 min
(9) Time for energizing the magnet per cycle = 0.5 min
(10) Labor required = 2 men per shift
(11) Amount of dispersant required = 10 ppm

Investment Costs

(1) Costs of major installed equipment ($)
    one HGMS unit $1,936,000
    pump   38,480
    tank   24,370
                                                        1,998,850
(2) Add 20% Contingency   399,770
    Total Investment (I $) $2,398,670

Operating Costs ($ per year)

(1) Dispersant (57¢/lb) $ 24,120
(2) Electric power (2¢/KWH, 650 KW)   102,960
(3) Water (3¢/1000 gal)   16,440
(4) Operating labor
    (2 men/shift X 8304 hr/yr X 6.5$/man-hr)   107,960
(5) Maintenance labor (1.5% of investment cost)   35,980
(6) Supervision (15% of operating and maintenance
    labor costs)   21,590
(7) Operating supplies (30% of operating labor cost)   32,390
(8) Maintenance supplies (1.5% of investment cost)   35,980
(9) Local taxes and insurance (2.7% of investment cost) 64,760
    Annual Net Operating Cost $\underline{N}$ $ $ 442,180
    Coal Processed Annually $\underline{G}$ tons   528,900

Unit Costs ($ per ton coal processed)
    See Ref. $\underline{29}$ and $\underline{34}$ for the cost equations used below.

(1) Based on 0% DCF rate of return:
    $\underline{U}_o = (\underline{N} + 0.05\ \underline{I})/\underline{G} = 1.063$ $/ton
(2) Based on 15% DCF rate of return:
    $\underline{U}_{15} = (\underline{N} + 0.34749\ \underline{I})/\underline{G} = 2.412$ $/ton
(3) Based on capital amortization over 20 years at 10% interest
    rate:
    $\underline{U} = (\underline{N} + 0.11746\ \underline{I})/\underline{G} = 1.369$ $/ton

Literature Cited

1. Slack, A.V., "Flue Gas Desulfurization An Overview," Chem. Eng. Progr., (1976), 72 (8), 94-97.
2. Davis, J.C., "Coal Cleaning Readies for Wider Sulfur-Removal Role", Chem. Eng. (March 1, 1976) 83, 70-74.
3. Kester, W.M., "The Effect of High-Intensity Magnetic Cleaning on Pulverized High Sulfur, Bituminous Coals", Master Thesis, School of Mines, West Virginia University, Morgantown, W. Va., 1966.
4. Lin, C.J., Liu, Y.A., Vives, D.L., Oak, M.J., Crow, G.E., "Sulfur and Ash Removal from Coals by High Gradient Magnetic Separation", ACS Div. Fuel Chem. Preprints (1976), 21 (7), 124 (Papers presented at 172nd National Meeting.)
5. Lin, C.J., Liu, Y.A., Vives. D.L., Oak, M.J., Crow, G.E., Huffman, E.L., "Pilot-Scale Studies of Sulfur and Ash Removal from Coals by High Gradient Magnetic Separation", IEEE Trans. Magn., (1976), MAG-12, 513-521.
6. Luborsky, F.E., "High Gradient Magnetic Separation for Removal of Sulfur from Coal", Final report, issued by the General Electric Corporate Research and Development to the Bureau of Mines under contract no. H0366008, February, 1977.
7. Oak, M.J., "Modeling and Experimental Study of High Gradient Magnetic Separation with Application to Coal Beneficiation", M.S. Thesis, Department of Chemical Engineering, Auburn University, Auburn, AL, 1977.
8. Trindade, S.C., "Studies on the Magnetic Demineralization of Coal", Ph.D. Thesis, Department of Chemical Engineering, Massachusetts Institute of Technology, Cambridge, MA, 1973.
9. Liu, Y.A., Editor, "Proceedings of Magnetic Desulfurization of Coal Symposium: A Symposium on the Theory and Applications of Magnetic Separation", IEEE Trans. Magn. (1976), MAG-12 (5), 423-551.
10. Liu, Y.A., Lin, C.J., "Assessment of Sulfur and Ash Removal from Coals by Magnetic Separation", IEEE Trans. Magn. (1976), MAG-12, 538-550.
11. Iannicelli, J., "Assessment of High Extraction Magnetic Filtration", Special Report to the National Science Foundation, available as document No. Pb240-880/5 from the National Technical Information Service, Springfield, VA, 1974.
12. Iannicelli, J., "New Developments in Magnetic Separation", IEEE Trans Magn. (1976), MAG-12, 436-443.
13. Liu, Y.A., Lin, C.J., Vives, D.L., Oak, M.J., Crow, G.E., "Theory and Applications of High Gradient Magnetic Separation: A Review", invited paper, International Magnetics Conference, Los Angeles, CA, 1977.
14. Luborsky, F.E., "High-Field High-Gradient Magnetic Separation: A Review", 21st National Conference on Magnetism and Magnetic Materials, Philadelphia, PA, December, 1975.
15. Oberteuffer, J.A., "Magnetic Separation: A Review of Prin-

ciples, Devices and Applications", IEEE Trans. Magn. (1974), MAG-10, 223-238.

16. Oberteuffer, J.A., "Engineering Development of High Gradient Magnetic Separation", IEEE Trans. Magn., (1976), MAG-12, 444-449.

17. Oder, R.R., "High Gradient Magnetic Separation: Theory and Applications", IEEE Trans. Magn. (1976), MAG-12, 428-435.

18. Liu, Y.A., Lin, C.J., "Status and Problems in the Development of High Gradient Magnetic Separation Processes Applied to Coal Beneficiation", in "Proceedings of Engineering Foundation Conference on Clean Combustion of Coal", Rindge, N.H., July 31-August 5, 1977; to be published by the National Technical Information Service, Springfield, VA, 1977.

19. Murray, H.H., "High Intensity Magnetic Cleaning of Bituminous Coals", National Coal Association and Bituminous Coal Research, Inc. Coal Conference and Expo III, Lexington, KY, 1976.

20. Cavallaro, J.A., Johnston, M.T., and Deurbrouck, A.W., "Sulfur Reduction Potential of the Coals of the United States", U.S. Bur. Mines, Rep. Invest. (1976), 8118.

21. Hoffman, L., "The Physical Desulfurization of Coal: Major Consideration of SO_2 Emission Control", Special Report, Mitre Corp., MeLean, VA, November, 1970.

22. Oder, R.R., "Pyritic Sulfur Removal from Coals by High Gradient Magnetic Separation", The 2nd National Conference on Water Re-Use, Chicago, IL, May, 1975.

23. Thompson, R.D., York, H.F., "The Reserve Base of U.S. Coals by Sulfur and Ash Content, Part I. The Eastern States", U.S. Bur. Mines, Inf. Circ. (1975), 8680.

24. Thompson, R.D., and York, H.F., "The Reserve Base of U.S. Coals by Sulfur and Ash Content, Part II. The Western States", U.S. Bur. Mines, Inf. Circ. (1975), 8693.

25. Crow, G.E., "A Pilot-Scale Study of High Gradient Magnetic Desulfurization of Solvent Refined Coal", M.S. Thesis, Department of Chemical Engineering, Auburn University, Auburn, AL 1977.

26. Hydrocarbon Research, Inc., Trenton, NJ, Report Nos. L-12-61-501, issued to Electric Power Research Institute of Palo Alto, under Contract RP-340, 1975.

27. Oder, R.R., "Magnetic Desulfurization of Liquefied Coals: Conceptual Process Design and Cost Estimation", IEEE Trans. Magn. (1976), MAG-12, 532-537.

28. Wolk, R., Stewart, N., Alpert, S., "Solvent Refining for Clean Coal Combustion", EPRI J., (May 1976), 12-16.

29. Batchelor, J.D., Shih, C., "Solid-Liquid Separation in Coal Liquefaction Processes", AIChE National Meeting, Los Angeles, CA, November, 1975.

30. Agarwal, J.C., Gilberti, R.A., Irminger, P.F., Sareen, S.S., Chemical Desulfurization of Coal", Min. Congr. J. (1975), 70 (3), 40-43.

31. Kindig, J.K., Turner, R.L., "Dry Chemical Process to Magnetize Pyrite and Ash Removal from Coal", Preprint No. 76-F-366, SME-AIME Fall Meeting, Denver, September, 1976.

32. Ouinlan, R.M., Venkatesan, S., "Economics of Coal Preparation", Preprint No. 76-F-108, AIME Annual Meeting, Las Vegas, NV, February, 1976.

33. Van Nice, L.J., Santy, M.J., Meyers, R.A., "Meyers Process: Plant Design, Economics and Energy Balance", National Coal Association and Bituminuous Coal Research, Inc. Coal Conference and Expo, III, Lexington, KY, October, 1976.

34. Federal Power Commission, "Final Report: The Supply-Technical Advisory Task Force on Synthetic Gas-Coal", April, 1973.

35. Guthrie, D.M., "Capital Cost Estimating", Chem. Eng. (March 24, 1969), 76, 114-142.

36. Iannicelli, J., personal communication, Aquafine Corp., Brunswick, GA, October, 1976.

Extraction of Sulfur from Coal by
Reaction and Leaching

Applicability of the Meyers Process for Desulfurization of U.S. Coal (A Survey of 35 Coals)

J. W. HAMERSMA, M. L. KRAFT, and R. A. MEYERS

TRW Systems and Energy, Redondo Beach, CA 90278

The Meyers process (1,2) is a new chemical leaching process which allows many coal-fired power plants and industrial sources to meet federal and state sulfur oxide emission standards without the use of flue gas cleaning. This process uses a regenerable aqueous ferric sulfate leaching unit to chemically convert and remove the pyritic sulfur content of the coal as elemental sulfur and iron sulfate. Although only pyritic sulfur is removed (organic sulfur remains), the Meyers process has wide applicability for converting U.S. coal reserves to a sulfur level consistent with present and proposed governmental sulfur oxide emission standards.

Thirty-five mines from the major coal basins were investigated in this study. Because of the relatively high pyritic sulfur and low organic sulfur contents and because of high production of Appalachian coals (70% of current U.S. output), the Meyers process appears to have major impact in this area.

In the Meyers process, aqueous ferric sulfate is used at 90°-130°C to oxidize selectively the pyritic sulfur content of coal to yield iron sulfate and free elemental sulfur as shown in Equation 1. The iron sulfate dissolves in solution while the free sulfur is removed from the coal matrix either by vaporization or solvent extraction. The leaching (oxidizing) agent is then regenerated at a similar temperature using oxygen or air and is recycled while product iron sulfates are removed by liming and/or crystallization.

$$4.6 \ Fe_2(SO_4)_3 + 4.8 \ H_2O + FeS_2 \rightarrow 10.2 \ FeSO_4 + 4.8 \ H_2SO_4 + 0.8 \ S \quad (1)$$

The detailed chemistry, leaching conditions, reaction kinetics, process engineering, and cost estimates have been published (3,4,5,6,7), and a reactor testing unit has been built.

This chapter presents experimental results and discussion for: pyritic sulfur removal from coal, pyritic sulfur partition by float-sink separation from the same coals, the effect of the Meyers process on the trace element content of the treated coals, and applicability of the Meyers process for meeting air pollution

control standards. This work was performed under contract to the
Environmental Protection Agency (8,9).

Sulfur Reduction

A summary of the sulfur analyses of the run-of-mine coal utilized in this study is shown in Figure 1. The organic and pyritic
sulfur contents are plotted along the x-and y-axis respectively
while the sum of these values for a given coal can be read from the
diagonal lines. The average coal for this survey contained 2.02%
pyritic sulfur and 3.05% total sulfur. This corresponds to the
U.S. Bureau of Mines average for 325 raw coals (10), indicating
that the survey coals represent reasonably the sulfur distribution
in U.S. coal.

The results to date for chemical removal of pyritic sulfur
(100-150 μ top-size coal) and the optimal results for conventional
coal washing (based on the 1.4 mm, 1.90 float fraction of a float-
sink analysis) are shown in Table I, in terms of total sulfur
changes, and in graphical form in Figure 2. The table describes
the results obtained on coals which contained sufficient pyritic
sulfur for accurate sulfur removal determination (i.e., >0.25% w/w).
Three of the mines sampled were below this limit and, therefore,
do not appear in the table. Actual total sulfur values before and
after chemical removal are shown in columns 3 and 4. These may
be compared with column 5, which shows sulfur values which can be
obtained with full process optimization. This latter value was
calculated by adding the actual residual pyritic sulfur and sulfate
contents of the coal to the initial organic sulfur value after
correction for any concentration effects resulting from ash removal. In the survey program, complete removal of residual elemental sulfur and sulfate was not always obtained since conditions
were standardized but not optimized for each individual coal.
Thus, for example, although 96% pyrite conversion was obtained for
the Bird No. 3 coal, the total sulfur was reduced to 0.80%, not
the theoretical 0.45% caused by these effects. These processing
problems have now been resolved as part of other projects (3,9),
and the values shown in column 5 can be considered to represent
the true potential of the process. Because of the widespread
application of physical cleaning techniques for removal of non-
combustible rock (which includes varying amounts of pyrite along
with some carbon) from coal, float-sink fractionation was performed in order to define the relative utility of washing and
chemical desulfurization for each coal. The results are shown in
column 8 and also in Figure 2. The sulfur reduction potential of
the Meyers process is highly attractive and in particular it was
found that:

    (1)   The Meyers process, at its current state of development,
            removed 83-99% of the pyritic sulfur content of the 32
            coals studied, resulting in total sulfur content re-
            ductions of 25-80%

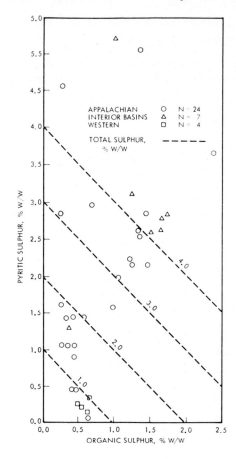

Figure 1. Sulfur forms of sampled U.S. coals

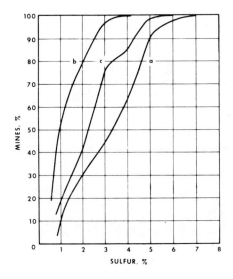

Figure 2. Sulfur content of survey run-of-mine coals (curve a), the same coals physically cleaned (curve c), and chemically desulfurized (curve b)

Table I.   Summary of Pyritic Sulfur Removal Results[1]
(100–150 μ top-size coal)

Mine	Seam	% Total Sulfur w/w in Coal[2]			Meyers Process		% Sulfur in Coal[3] After Float-Sink
		Initial	After Meyers Process Current Results	Revised Processing[4]	Pyrite Conv. (% w/w)	Total S Decrease (% w/w)	
Navajo	Nos. 6,7,8	0.8	0.6	0.5	90	25	---
Kopperston No. 2	Campbell Creek	0.9	0.6	0.5	92	33	0.8
Harris Nos. 1 & 2	Eagle & No. 2 Gas	1.0	0.8	0.6	94	23	0.9
Colstrip	Rosebud	1.0	0.6	0.7	83	30	---
Warwick	Sewickley	1.4	0.6	0.4	92	54	1.0
Marion	Upper Freeport	1.4	0.7	0.6	96	50	1.2
Mathies	Pittsburgh	1.5	0.9	0.5	95	36	1.7
Isabella	Pittsburgh	1.6	0.7	0.5	96	54	1.5
Orient No. 6	Herrin No. 6	1.7	0.9	0.8	96	44	1.4
Lucas	Middle Kittanning	1.8	0.6	0.5	94	64	0.7
Jane	Lower Freeport	1.8	0.7	0.6	91	63	0.8
Martinka	Lower Kittanning	2.0	0.6	0.8	92	70	0.8
North River	Corona	2.1	0.9	0.8	91	55	2.2
Humphrey No. 7	Pittsburgh	2.6	1.5	1.2	91	42	1.9
No. 1	Mason	3.1	1.6	1.4	90	48	2.3
Bird No. 3	Lower Kittanning	3.1	0.8	0.4	96	75	1.5
Williams	Pittsburgh	3.5	1.7	1.4	96	50	2.3
Shoemaker	Pittsburgh	3.5	1.7	1.5	96	51	3.6
Meigs	Clarion 4A	3.7	1.9	1.7	93	48	2.8
Fox	Lower Kittanning	3.8	1.6	1.2	89	57	2.0
Dean	Dean	4.1	2.1	1.6	94	49	3.0
Powhattan No. 4	Pittsburgh No. 8	4.1	1.9	1.9	85	53	3.3
Eagle No. 2	Illinois No. 5	4.3	2.0	1.9	94	54	2.9
Star	No. 9	4.3	2.5	1.8	91	43	3.0
Robinson Run	Pittsburgh	4.4	2.2	1.6	97	50	3.0

Table I continued...

| Mine | Seam | % Total Sulfur w/w in Coal[2] | | | Meyers Process | | % Sulfur in Coal[3] |
| | | Initial | After Meyers Process | | Pyrite Conv. (% w/w) | Total S Decrease (% w/w) | After Float-Sink |
			Current Results	Revised Processing[4]			
Homestead	No. 11	4.5	1.7	1.6	93	47	3.2
Camp Nos. 1 & 2	No. 9 (W. Ky.)	4.5	2.0	2.1	89	55	2.9
Ken	No. 9	4.8	2.8	2.2	91	42	3.5
Delmont	Upper Freeport	4.9	0.8	0.5	96[5]	80	2.1
Muskingum	Meigs Creek	6.1	3.2	2.7	94	47	4.4
Weldon No. 11	Des Moines No. 1	6.4	2.2	1.7	92	65	3.9
Egypt Valley No. 21	Pittsburgh No. 8	6.6	2.7	2.1	89	59	4.6

[1] A 100 g sample of coal was refluxed with 2 L of $1\underline{N}$ $Fe_2(SO_4)_3$ solution. After 4–6 hrs, the reduced solution was filtered from the coal and the coal was refluxed with 2 L of fresh $1\underline{N}$ $Fe_2(SO_4)_3$. After a total extraction time of 10–24 hr, the slurry was filtered and the coal cake was washed with $0.2\underline{N}$ H_2SO_4, then with water to remove residual sulfate. Toluene (400 ml) extraction of the coal removed elemental sulfur. The coal cake was then dried in a vacuum oven at 100–120°C.

[2] Dry, moisture-free basis.

[3] 1.90 float material, 14 mesh x 0, is defined here as the limit of conventional coal cleaning.

[4] Calculated using latest process improvement.

[5] Run at 200 m x 0.

(2) Twelve (38%) of the coals were reduced in sulfur content
     to the 0.6-0.8% sulfur levels generally consistent with
     the Federal New Source Performance Standards and many
     state standards
(3) In all cases, the Meyers process removed significant-to-
     very large increments of sulfur over that separable by
     physical cleaning
(4) In two cases, the Corona and Mathies mines, coal clean-
     ing actually resulted in a sulfur content increase in
     the float product.

Rate of Pyritic Sulfur Removal

The removal of pyritic sulfur was measured as a function of
time at 100°C for 18 Appalachian and 3 eastern interior region
coals. The results are displayed in Table II, which shows the
range of rates observed. It was assumed that the empirical kinetic rate expression (3) which was developed previously for this
process is applicable to all coals in the survey. The kinetic
equation can be simplified by holding the reagent concentration
relatively constant, as was the case in this study, to yield
Equation 2:

$$\frac{-d[W_p]}{dt} = k^o W_p^2 = \text{rate of pyrite removal} \tag{2}$$

where: W_p = weight percent pyrite in the coal, and $\underline{k}^o$ = function
of temperature, coal type and partical size.
By integrating Equation 2, the fraction of pyrite removed as a
function of time is shown in Equation 3:

$$\frac{F}{1-F} = k^o W_p^o t_F \tag{3}$$

where: $\underline{F}$ - fraction of pyrite removed, W_p^o = initial pyrite concentration, and $\underline{t}_F$ = time to removal at fraction F.
The initial weight percent of pyritic sulfur $\underline{Sp}^o$ may be substitued
for W_p^o and Equation 3 be rearranged to Equation 4:

$$\frac{1}{Sp^o t_F} \; \alpha \; \underline{k} = \text{actual rate constant} \tag{4}$$

Thus, assuming 80% removal as a point of comparison, the values of
$1/Sp^o t_{80\%}$ shown in column 6 of Table II indicate the reactivities
of the pyrite contained in the coals that were studied. A large
amount of experimentation and engineering has been performed using
rate data obtained for Martinka (3) coal; therefore, the Martinka
coal $Sp^o t_{80\%}$ has been set equal to 1 as a basis of comparison (as
shown in column 7).

Table II. Relative Rate Constants for Pyritic Sulfur Removal[1]

Coal Mine	Seam	Top Size (Microns)	Sp^o	$t_{80\%}$ (hrs)	Realtive Rate Constants $\frac{1}{Sp^o t_{80\%}}$	Relative[2] Rate
Kopperston No. 2	Campbell Creek	149	0.47	2.0	1.1	16
Harris Nos. 1 & 2	Eagle & No. 2 Gas	149	0.49	2.3	0.89	13
Marion	Upper Freeport	149	0.90	3.0	0.37	5.3
Lucas	Middle Kittanning	100	1.42	3.25	0.22	3.1
Shoemaker	Pittsburgh	149	2.19	2.9	0.16	2.3
Williams	Pittsburgh	100	2.23	3.0	0.15	2.1
Ken	No. 9	149	2.85	2.5	0.14	2.0
North River	Corona	149	1.42	5.0	0.14	2.0
Star	No. 9	100	2.66	3.0	0.13	1.9
Mathies	Pittsburgh	100	1.05	9.0	0.11	1.6
Powhattan No. 4	Pittsburgh No. 8	75	2.75	4.0	0.091	1.3
Homestead	No. 11	149	3.11	3.5	0.092	1.3
Fox	Lower Kittanning	75	3.09	4.5	0.072	1.0
Isabella	Pittsburgh	149	1.07	13.0	0.072	1.0
Martinka	Lower Kittanning	149	1.42	10.0	0.070	1.0
Meigs	Clarion 4A	149	2.19	8.5	0.054	0.77
Bird No. 3	Lower Kittanning	100	2.87	8.0	0.044	0.62
Dean	Dean	100	2.62	10.2	0.037	0.53
Muskingum	Meigs Creek No. 9	149	3.65	8.0	0.034	0.49

1 As in footnote 1 of Table I, except toluene extraction was omitted. Pyrite removal rate data was obtained by pyrite analysis of coal samples obtained from slurry aliquots obtained by centrifugation.

2 $1/Sp^o t_{80\%}$ relative to value for Martinka mine.

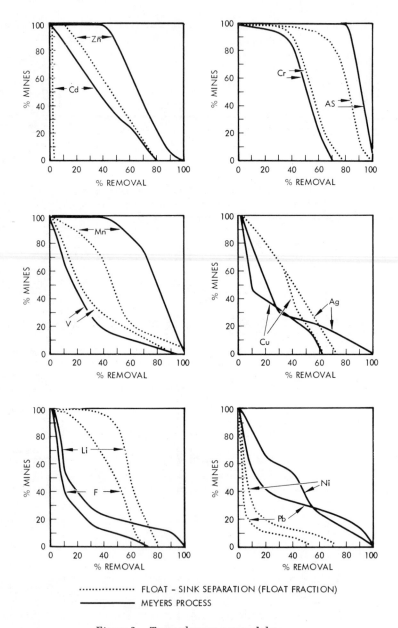

Figure 3.   Trace element removal data

It can be easily seen from Table II that there is a wide band of rate constants rather evenly spread over a factor of approximately 30. The Kopperston No. 2 and Harris Nos. 1 and 2 coals react more rapidly than the slowest coals (Dean and Muskingum) by a factor of about 30. Thus, it is apparent that real and significant rate differences do exist between pyrite in various coals. Characteristics of coal such as pore structure, size and shape distribution of pyrite, etc. may be the primary factors affecting the rate constant as reflected in the observed band of values found for the rates given in Table II.

Trace Element Removals

Because both chemical leaching and physical cleaning processes can remove potentially harmful trace elements from coal either by leaching or by physical partitioning, 50 coal samples have been analyzed in duplicate or triplicate to determine the extent of the removal, if any, of 18 elements of interest to the Environmental Protection Agency. The samples included 20 as received, 20 chemically leached, and 10 float-sink treated coal samples. The results are shown for 12 elements in Figure 3 in cumulative fashion. The extraction procedure was as in Table I. Six elements, B, Be, Hg, Sb, Se, and Sn yielded negative or inconclusive results because of low levels or analysis difficulties and thus are not plotted. Although the results varied greatly from coal to coal with respect to the elements extracted and the degree of extraction, some general conclusions can be reached.

(1) As, Cd, Mn, Ni, Pb, and Zn are removed to a significantly greater extent by the Meyers process
(2) F and Li are partitioned to a greater extent by physical separation procedures.
(3) Ag and Cu are removed with a slight preference for float-sink separation
(4) Cr and V are removed by both processes with equal success.

The effective removal of As, Cd, Cr, Ni, Pb, and Zn from the coal by the Meyers process is especially noteworthy as these compounds are concentrated (along with Se) in the fine particulates emitted from coal-fired power plants. This fine particulate matter passes through conventional fly-ash control devices.

Literature Cited

1. Meyers, R.A., Hamersma, J.W., Land, J.S., Kraft, M.L., Science, (1972), 177, 1187.
2. Meyers, R.A., "Removal of Pyritic Sulfur from Coal Using Solutions Containing Ferric Ions," U.S. Patent 3768988, 1973.
3. Koutsoukos, E.P., Kraft, M.L., Orsini, R.A., Meyers, R.A., Santy, M.J., Van Nice, L.J., "Final Report Program for Bench-Scale Development of Processes for the Chemical Extraction of

Sulfur from Coal," <u>Environ. Prot. Technol. Ser.</u>, (May 1976),
EPA-600/2-76-143a.

4. Hamersma, J.W., Koutsoukos, E.P., Kraft, M.L., Meyers, R.A.,
   Ogle, G.J., Van Nice, L.J., "Chemical Desulfurization of Coal:
   Report of Bench-Scale Developments, Volumes 1 and 2,"
   <u>Environ. Prot. Technol. Ser.</u>, (February 1973),EPA-R2-173a.

5. Magee, E.M., "Evaluation of Pollution Control in Fossil Fuel
   Conversion Processes, Coal Treatment: Section 1. Meyers
   Process," <u>Environ. Prot. Technol. Ser.</u>, (September 1975),
   EPA-650/2-74-009k.

6. Nekervis, W.F., Hensley, E.F., "Conceptual Design of a Commercial Scale Plant for Chemical Desulfurization of Coal,"
   <u>Environ. Prot. Technol. Ser.</u> (September 1975), EPA-600/2-75-051.

7. Tek, M. Rasin, "Coal Beneficiation," in Evaluation of Coal
   Conversion Processes, (1974) <u>PB-234202</u>.

8. U.S. Environmental Protection Agency, Office of Research and
   Development, Washington, DC, "Applicability of the Meyers
   Process for Chemical Desulfurization of Coal: Initial Survey
   of Fifteen Coals," by J.W. Hamersma <u>et al</u>, Systems Group of
   TRW, Inc., Redondo Beach, CA. (April 1974),EPA-650/2-74-025,
   Contract No. 68-02-0647.

9. Hamersma, J.W., Kraft, M.L., "Applicability of the Meyers
   Process for Chemical Desulfurization of Coal: Survey of
   Thirty-Five Coals," <u>Environ. Prot. Technol. Ser.</u>, (September
   1975), EPA-650/2-74-025a.

10. Deurbrouck, A.W., "Sulfur Reduction Potential of Coals of the
    United States," <u>U.S. Bur. Mines Rep. Invest</u>. (1972), <u>7633</u>.

Coal Desulfurization Test Plant Status—July 1977

L. J. VAN NICE, M. J. SANTY, E. P. KOUTSOUKOS, R. A. ORSINI, and R. A. MEYERS

TRW Systems and Energy, Redondo Beach, CA 90278

An 8-metric ton/day process test plant for chemical desulfurization of coal utilizing ferric sulfate leach solution, has just been built at TRW's Capistrano Test Site in California. The plant, shown in Figures 1 and 2, was constructed under an Environmental Protection Agency sponsored project for the development of the Meyers process. Current plans call for plant shakedown followed by processing of 100-200 tons of American Electric Power Service Corporation's Martinka mine coal. The Meyers process and its applicability for directly desulfurizing raw run-of-mine coal were presented in the previous chapter and the process is described in connection with the broad field of chemical desulfurization in a newly issued book (1).

We have very recently verified that the ferric sulfate leach solution, which has a specific gravity of 1.2-1.4, can be used to gravity separate coal into float and sink fractions in a manner very advantageous to the Meyers process. This float-sink approach was tested many years ago by A. Z. Yurovskii in the U.S.S.R but only published in the open literature very recently (2). Yurovskii subsequently treated the sink coal with a mixture of nitric acid and ferric sulfate to remove pyritic sulfur, not realizing, that the separation medium itself was sufficiently active to accomplish near total pyritic sulfur removal.

In fact, a practical method for actual float-sink cleaning of coal in a dense liquid has long been sought as an alternative to mechanical cleaning. Heavy liquids, such as zinc chloride-water, or chlorinated, brominated, or fluorinated hydrocarbons are useful for prediction of yields which can theoretically be obtained in mechanical washing plants, but are impractical for actual production because they are expensive and add pollutants to the coal, atmosphere or water table. We find that the density of aqueous ferric sulfate leach solution, as utilized in the Meyers process, is ideal for accomplishing a practical gravity separation of coal for specific gravities between 1.2 and 1.4. For about 40% of Appalachian coal, the float coal (which is often 40-60% by weight of the total) averages 0.8-1.2 lbs $SO_2/10^6$ btu and needs

Figure 1. Test plant—front view

Figure 2. Test plant—view through tank farm

no further processing to meet standards, while the sink coal
slurry contains most of the coal pyrite. This coal slurry can be
processed through the Meyers process to give coal containing 1.2-
1.5 lbs $SO_2/10^6$ btu. This approach allows production of coal which
will meet air pollution control standards for both New and Existing Stationary Sources, and reduces the cost of the Meyers process
through allowing 40-60% of the coal to bypass reactor, elemental
sulfur extraction and dryer units.

We term this approach, Gravichem Cleaning. Where estimated
processing costs using utility financed depreciation of capital
and including coal grinding and compaction were $10-12/ton (1),
forecasts for the new Gravichem Cleaning approach are $8.50/ton
($0.35/10^6$ btu).

Process Chemistry, Kinetics and Scheme

The process is based on the oxidation of coal pyrite with
ferric sulfate (Equation 1). The leaching reaction is highly
selective to pyrite with 60% of the pyritic sulfur converted to
sulfate sulfur and 40% to elemental sulfur. The reduced ferric
ion is regenerated by oxygen or air according to Equations 2 or 3.

$$FeS_2 + 4.6\ Fe_2(SO_4)_3 + 4.8\ H_2O \rightarrow 10.2\ FeSO_4 + 4.8\ H_2SO_4 + 0.8\ S \quad (1)$$

$$2.4\ O_2 + 9.6\ FeSO_4 + 4.8\ H_2SO_4 \rightarrow 4.8\ Fe_2(SO_4)_3 + 4.8\ H_2O \quad (2)$$

$$2.3\ O_2 + 9.2\ FeSO_4 + 4.6\ H_2SO_4 \rightarrow 4.6\ Fe_2(SO_4)_3 + 4.6\ H_2O \quad (3)$$

Regeneration can be performed either concurrently with coal
pyrite leaching in a single operation or separately. The net
effect of the process is the oxidation of pyrite with oxygen
to yield recoverable iron, sulfate sulfur, and elemental sulfur.
The form of process products varies to some extent with the degree
of regeneration. Thus, Equations 1 and 2 lead to the overall process chemistry indicated by Equation 4 producing a mixture of iron
sulfates and elemental sulfur. Equations 1 and 3 yield ferrous
sulfate, sulfuric acid, and elemental sulfur as indicated by
Equation 5.

$$FeS_2 + 2.4\ O_2 \rightarrow 0.6\ FeSO_4 + 0.2\ Fe_2(SO_4)_3 + 0.8\ S \quad (4)$$

$$FeS_2 + 2.3\ O_2 + 0.2\ H_2O \rightarrow FeSO_4 + 0.2\ H_2SO_4 + 0.8\ S \quad (5)$$

Several options exist in product recovery. Iron sulfates may
be recovered as pure solids by stepwise evaporation of a spent reagent slipstream with ferrous sulfate being recovered first because of its lower solubility. Alternately, ferrous sulfate may
be recovered by crystallization and ferric sulfate or sulfuric

acid removed by liming spent reagent or spent wash water slip-
streams. Iron sulfates may be stored as solids for sale or may be
converted easily to highly insoluble basic iron sulfates (by air
oxidation) or calcium sulfate (by low-temperature solid phase
reaction) for disposal. Elemental sulfur may be recovered from
coal by vaporization with steam or by vacuum, or it can be leached
out with organic solvents such as acetone depending on product
marketability and product recovery economics. Recovery economics
may be influenced by quantity and concentration of product in the
process effluent streams which in turn are influenced by the pyrite
concentration in the coal and the desired extent of desulfurization.

The process has been extensively studied at bench-scale.
Parameters investigated include coal top-size, reagent composition,
slurry concentration, reaction temperature and pressure, and reaction time.

Additional investigations completed or underway include concurrent coal leaching-reagent regeneration, product recovery,
product stability, and the effect of coal physical cleaning on
process performance and economics. The process scheme depicted
in Figure 3 is based on the bench-scale testing. Coal is
  (1) Crushed to the desired size for processing
  (2) Contacted with hot recycled reagent in the mixer
       (90°-100°C)
  (3) Leached of pyrite in the reactor(s) with simultaneous
       or separate reagent regeneration
  (4) Washed with hot water
  (5) Stripped of elemental sulfur, dried and finally cooled.
The iron and sulfate sulfur are recovered from spent reagent slip-
streams prior to reagent recycle. Figure 4 shows typical data on
pyrite removal rates from Appalachian coal as a function of temperature. During slurry mixing and heat-up, 10-20% of the pyrite
is removed.

Bench-scale data indicated that the pyrite leaching rate from
coal can be adequately represented by the empirical rate expression (Equation 6).

$$r_L = - \frac{dW_p}{dt} = K_L \, W_p^2 \, Y^2 \tag{6}$$

where: $K_L = A_L \exp(-E_L/RT)$, W_p = wt% pyrite in coal, Y = ferric
ion-to-total iron ratio in the reactor reagent, and A_L and E_L are
constants for each coal and particle size at least over most of
the reaction range.

The leach rate is a function of coal type. Pyrite extraction
rates vary considerably, as detailed in a study of the Meyers
process as applied to U.S. coal (3), e.g., there was more than one
order of magnitude difference between the fastest and slowest
reacting coal. The reagent regeneration rate is governed by the
rate expression (Equation 7).

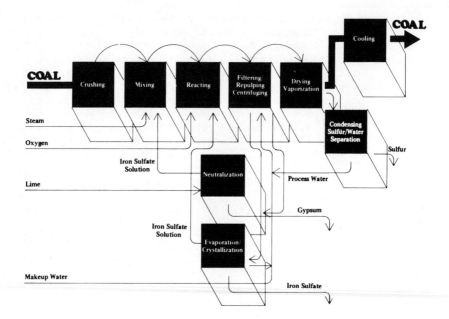

Figure 3.   Process flow schematic

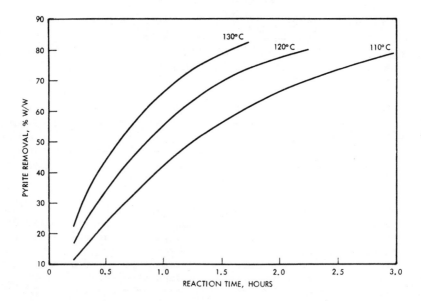

Figure 4.   Temperature effect on processing of 14 mesh top-size Lower Kittanning coal (33% w/w slurries)

$$r_R = - \frac{dFe^{+2}}{dt} = K_R P_{O_2} (Fe^{+2})^2 \qquad (7)$$

where: $K_R = A_R \exp(-E_R/RT)$, P_{O_2} = oxygen partial pressure, Fe^{+2} = ferrous ion concentration in the reagent solution, and A_R and E_R are constants.

Engineering evaluation of available data shows that it is preferable to process fine coal (<2 mm top-size) under simultaneous leaching-regeneration conditions at 110°-130°C until most of the pyrite is leached out. Ambient pressure processing (approximately 100°C) is indicated to remove the last few tenths percent of pyrite since the low W_p value substantially reduces the rate of ferric ion consumption and, therefore, the need for simultaneous reagent regeneration. Ambient pressure processing appears to be indicated also for coarse coal (e.g., 10 mm top-size) for several reasons. It is difficult to feed continuously a non-slurryable coal into and remove it from a pressure vessel. It is much easier and less costly to drain leach solution from the coal and pump it into a small pressure vessel for regeneration. Also the slower reaction rate with coarse coal would require much longer residence times and unreasonably large total volume for pressure vessels. These engineering evaluations were part of the data used to design the test plant.

Test Plant Design and Operation

A test plant sized to process up to 8 metric tons per day of coal has been built under the sponsorship of the Environmental Protection Agency at TRW's Capistrano test site. A plant flow diagram is shown in Figure 5. The facility can evaluate the following critical on-line operations:

(1) Pressure leaching of pyritic sulfur from 150 μ to 2 mm top-size coal at pressures up to 100 psig
(2) Regeneration of ferric sulfate both separately, for processing larger top-size coal or low pyrite coal, and in a single vessel with the leaching step for processing of suspendable coal
(3) Filtration of leach solution from reacted coal
(4) Washing of residual iron sulfate from the coal

Iron sulfate crystallization, elemental sulfur recovery, and coal-drying unit operations will be evaluated in an off-line mode in equipment vendor pilot units. Leaching of 10 mm top-size coal can be evaluated in an off-line mode in an atmospheric pressure vessel in the test plant. Coarse coal processing (5-10 mm top-size) has been very promising in laboratory tests (4). If this approach proves out in bench-scale evaluations, more extensive and on-line coal leaching units can be added readily to the present test plant. Processing fine coal allows the highest rate of pyritic sulfur

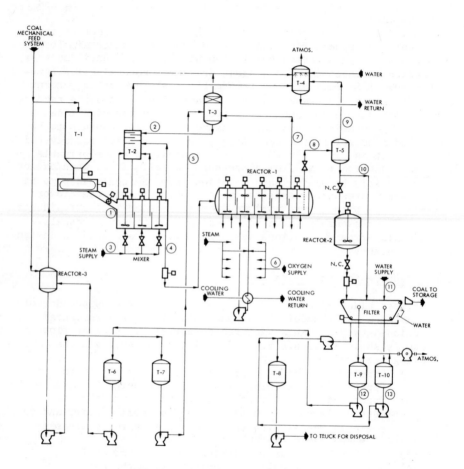

Figure 5.   Test plant flow diagram

removal while processing coarse coal, although slower, allows
lower cost coal dewatering units and the direct shipping of desulfurized coal product without need for pelletizing.

The test plant constructed at the Capistrano test site is a
highly flexible facility capable of testing the numerous alternate
processing modes of potential interest in the Meyers process. The
flow diagram shown in Figure 5 presents an equipment train for continuous process testing of slurried coal. Fine coal ground to the
desired size is stored under nitrogen gas in 1.8 metric ton sealed
bins. As required, bins are emptied into the feed tank (T-1).
Dry coal is fed continuously by a live bottom feeder to a weigh
belt which discharges through a rotary valve to the three stage
mixer (stream 1). The aqueous iron sulfate leach solution (stream
2) enters the mixer after first passing through a foam breaker
(T-2). Steam is added (stream 3) to raise the slurry to its boiling point. Foaming occurs in the early stages of mixing, but
ceases when particle wetting is complete. It is believed that the
mixing time and conditions necessary to complete the wetting and
defoaming of the slurry depend on the coal type and size and on
the residual moisture in the feed coal. To allow study of the
mixing parameters, the mixer stages have variable volume with
variable speed agitators, and the feed flow rates for coal, leach
solution and steam can be varied over wide ranges.

The defoamed slurry (stream 4) is pumped to a five-stage
pressure vessel (reactor 1) in which most of the pyrite is removed.
Some of the pyrite reaction occurs during mixing, but in the mixer
the reaction rate slows rapidly because the remaining pyrite ($\underline{W}_p$)
decreases and because the ferric iron is rapidly being converted
to ferrous iron ($\underline{Y}$ decreases). The pressure reactor overcomes the
decreased rate in two ways. First, it increases the temperature
(and pressure) to increase the reaction rate constant. Second,
oxygen is introduced under pressure to regenerate ferric iron and
to maintain a high solution $\underline{Y}$. The flow diagram shows that steam
and oxygen can be added to any or all of the five stages and that
cooling can be provided for any stage if necessary to remove the
excess heat of reaction. The unused oxygen saturated with steam
(stream 7) is contacted in a small pressure vessel (T-3) with the
feed leach solution (stream 5) to provide heated leach solution
for the mixer (stream 2) and cooled vent gas. The vent gas from
both T-2 and T-3 are scrubbed in T-4 to remove any traces of acid
mist. The important reaction parameters have already been well
studied at laboratory and bench-scale in batch mode. The test
plant reactor will accommodate the necessary studies of key parameters in a continuous reactor at coal throughputs between 2 and
8 metric tons per day. Parameters which will be studied include
temperature, pressure, oxygen purity, slurry concentration, iron
sulfate concentration, ferric-to-total iron ratio, acid concentration, residence time per stage, number of stages, mixing energy,
type of mixing, coal size, and type. The reactor can also be used
to study leach solution regeneration in the absence of coal.

Reacted coal slurry (stream 8) at elevated temperature and pressure is flashed into a gas-liquid separator vessel (T-5). The steam generated (stream 9) is condensed in T-4, and the condensate plus any entrained acid mist is removed with the water. The residual slurry (stream 10) is fed to a belt filter. The filtrate, which is regenerated leach solution, is removed from the coal slurry through a vacuum receiver (T-9) and pumped (stream 12) to a large leach solution storage tank (T-6). The coal on the filter belt is washed with water (stream 11) and discharged from the filter belt. The wash water is removed through a vacuum receiver (T-10) and sent to a large liquid waste holding tank (T-8) for subsequent disposal. The filter is a highly versatile unit which should provide the data necessary for scale-up. It has variable belt speed, variable belt areas assigned to washing, variable cake washing rates, belt sprays if needed to control blinding of the pores in the belt, and steam nozzles to provide for partial cake drying.

As an alternate process step, the slurry from the flash tank (T-5) can be passed into a secondary reaction vessel (reactor 2). At typical coal feed rates, this vessel can be filled in about 2 hr and then closed off, stirred and heated for any desired time before being pumped to the filter. Residence times up to about 10 hr are available in the primary reactor (reactor 1). This secondary reactor can be used to extend residence times to much longer times for examining the removal of final traces of pyrite or examining any other long-term behavior. The stirred vessel also can repulp the filter cake for additional coal washing studies.

The final item of major equipment in the test plant is the coarse coal contact vessel (reactor 3). This insulated and heated tank will hold a full bin (about 1.8 metric tons) of coarse coal (5-10 mm top-size). This vessel is used principally to convert the regenerated leach solution in storage tank T-6 to a more depleted solution in the process feed tank, T-7. In general, the iron sulfate leach solution in the filtrate going to tank T-6 will have a high Y because no secondary reactor was in use. For some test conditions, the feed to the process must be at a lower Y to simulate recycle leach solution from a secondary reactor. Passing all or some portion of the solution through coal will lower the Y of the solution to the desired value. This vessel is basically a coarse coal reactor, and if appropriate sampling ports and possibly some flow distribution internals were added, it could be used to obtain design data for coarse coal processing.

Solution tanks are sized at about 50,000 L to provide for about a week of continuous operation on the same feed without recycle or change. It also provides for uniform leach solution and coal samples of a large enough size for product recovery studies performed by equipment vendors. Operation at the scale of the test plant will provide adequate experience and data for the design of a demonstration-size commercial plant.

Literature Cited

1. Meyers, R.A., "Coal Desulfurization," Marcel Dekker, Inc., New York (1977).
2. Yurovskii, A.Z., "Sulfur in Coals," U.S. Department of Commerce, National Technical Information Service, Springield, Va. 22161 (1974).
3. Kraft, M.L., Hamersma, J.W., "Applicability of the Meyers Process for Chemical Desulfurization of Coal: Survey of 35 Coal Mines," Environ. Prot. Technol. Ser., (September 1975), EPA-650/2-74-025-a.
4. Koutsoukos, E.P., Kraft, M.L., Orsini, R.A., Meyers, R.A., Santy, M.J., Van Nice, L.J., "Final Report Program for Bench-Scale Delevopment of Processes for the Chemical Extraction of Sulfur from Coal," Environ. Prot. Technol. Ser., (May 1976), EPA-600/2-76-143a.

13

Oxidative Desulfurization of Coal

SIDNEY FRIEDMAN, ROBERT B. LACOUNT, and ROBERT P. WARZINSKI

Pittsburgh Energy Research Center, U.S. Energy Research and Development
Administration, 4800 Forbes Ave., Pittsburgh, PA 15213

It is becoming increasingly apparent that the solution to
our national energy problems will require a variety of approaches
and that these must be compatible with environmental restrictions. Coal, only recently considered destined for obscurity,
has been rescued by a combination of international political
events and increasing difficulties in developing a nuclear
power industry. Although coal as an energy source presents
problems, at least it is available and can be utilized.

The Federal Government, as part of the program administered
by the Energy Research and Development Administration, is carrying out research on many phases of coal utilization to overcome
the environmental problems involved in the combustion of coal.
One research project, which has been in progress at the
Pittsburgh Energy Research Center since 1970, is concerned with
chemical beneficiation of coal and most specifically with removal of sulfur from coal prior to combustion.

Experimental

Batch Experiments. Thirty-five grams of -200 mesh coal and
100 ml water were placed in a liner (glass or teflon) in a 1-L,
magnetically stirred, stainless steel autoclave. The autoclave
was pressurized with cylinder air to the required gauge pressure
(800 psi in the experiments in Tables I, II, and III) and then
was heated with stirring at 800–1000 rpm until the specified
temperature was reached (approximately 1 hr heat-up time).
Operating pressure for 800 psi initial experiments was 1000–1200
psi, depending on temperature and extent of reaction. After a
specified time at reaction temperature, ranging from 5 min to 2
hr, the autoclave was cooled by an internal cooling coil. Final
pressure at room temperature was between 650 and 750 psi. The
contents were removed, filtered, washed until the pH of the
filtrate was neutral, and then extracted in a Soxhlet thimble
with water until sulfate (present as $CaSO_4$) was no longer present
in the fresh extract. The coal was then dried thoroughly in a

vacuum oven at 100°C and analyzed by the Coal Analysis Section, U.S. Bureau of Mines.

   Semicontinuous Experiments. Using a similar autoclave fitted with pressure regulating valves, the autoclave containing the coal and water was heated to the specified temperature under 1 atm (initial pressure) N_2. At temperature, or shortly before reaching it, air was admitted to the desired pressure, which was 1000 psi except for experiments designed to investigate the effect of pressure.

   Temperature was kept at the required value by heating and cooling (cooling coil) while air (approximately 2 cuft/hr) flowed through the autoclave. After the required time at temperature, the autoclave was cooled, and the products were worked up as in the previous example.

Results and Discussion

   Although the project was initially divided into removal of organic and of inorganic sulfur (1), it was soon evident that, though one could remove pyritic sulfur without removing organic sulfur, the reverse was not true. Any process which removed organic sulfur would also remove pyritic sulfur. So the approach to the problem became one of finding chemical reactions suitable for removing organic sulfur from coal.

   The chemistry which we chose to explore was based on two premises: the major portion of the organic sulfur in coal was of the dibenzothiophene (DBT) type, and, the reagents had to be inexpensive.

   While we now believe that at least a sizable fraction of the organic sulfur in coal is not dibenzothiophenic, we have no reason to doubt that over 50% of it may be.

   These premises led us to the following hypothetical two-step removal of organic sulfur from coal.

   (1). Oxidation of organic (or dibenzothiophenic) sulfur to sulfone.

$$(1)$$

   (2). Elimination of the SO_2 from sulfone by base.

$$(2)$$

Both of these reactions are in the literature, so we had to modify and improve them so that they could be applied to desulfurization of coal.

The second step -- the removal of SO_2 from DBT sulfone by base -- was essentially quantitative when the sulfone was heated to 300°C in the presence of aqueous NaOH and nearly as efficient with Na_2CO_3. This was an improvement on the nonaqueous treatment (2).

The first step in the reaction -- oxidation to sulfone -- although extensively documented in the literature, presented more of a challenge. There are numerous oxidants reported which can effect the conversion of organosulfur compounds to sulfones, including $KMnO_4$, HNO_3, CrO_3, H_2O_2/HOAc, and hydroperoxides (3). These obviously do not fit the second premise that the reagents must be inexpensive. It was agreed that the only reagent which could be used as an oxidant was the oxygen in air. But DBT, and presumably the organic sulfur in coal, is inert to air at relatively high pressure and temperature. Transfer of oxygen to a carrier to form a hydroperoxide, followed by reaction of the hydroperoxide with DBT, did give sulfone. We found that with many hydrocarbons, such as tetralin, decalin, and cyclohexane, merely heating DBT with air under pressure in the presence of the hydrocarbon resulted in formation of sulfone (4), presumably as a result of in situ formation of hydroperoxides. Benzene, which does not form a hydroperoxide, affords no sulfone formation under comparable conditions.

Applying our two-step reaction -- air oxidation followed by treatment with aqueous base -- to coal, we removed up to 50% of organic sulfur, and almost completely eliminated pyritic sulfur as a bonus. Although this scheme appeared promising, it did require a suitable organic liquid and also NaOH.

We also explored another oxidation system which uses air as the ultimate source of oxygen. Nitrogen dioxide (NO_2) is a good reagent for converting sulfides to sulfones, and it can be used in an easily regenerable system.

$$2NO_2 + -S- \rightarrow 2NO + -SO_2- \qquad (3)$$

$$2NO + O_2 \rightarrow 2NO_2 \qquad (4)$$

We found that we could indeed oxidize DBT to its sulfone using NO_2 and air. When the reaction was extended to coal, however, a significant amount of concurrent reaction took place, including nitration of the coal, which consumed the nitrogen oxides and thus would have necessitated a continuous addition of NO_2 rather than the recycling shown in Equations 3 and 4.

In the meantime, our experiments on air oxidation of organosulfur using hydroperoxide precursors led us to the ultimate experiment, the one in which H_2O was used in place of an organic liquid phase. This reaction of coal with water and compressed air almost quantitatively converted the pyritic sulfur in coal to H_2SO_4. In addition, we also removed 25% of the organic sulfur. Here was evidence that there was some organosulfur in coal which was not DBT-like, since DBT failed to react with air and water under these conditions.

Initial experiments on the air-water oxydesulfurization of coal were carried out using a batch, stirred autoclave system. In this apparatus, in order to replace oxygen as it was used, it was necessary to cool the autoclave to near room temperature, vent the spent air, repressure, and reheat. Although this gave satisfactory desulfurization, it was an impractical approach for studying reaction parameters. The results cited in Tables I, II, and III are from 1-hr batch studies without repressurization and thus represent less than maximum desulfurization in some cases. All coals, except those noted, are bituminous.

The apparatus has been modified to allow air to flow through the stirred reactor while the coal-water slurry remains as a batch reactant. This is our current system. In this way, we are studying many of the variables as they will affect the reaction in a continuous system.

Our newest apparatus, now beginning operation, is a fully continuous unit, feeding both air and coal-water slurry into a reactor tube. This sytem is designed to obtain data on reaction rates, develop information for economic evaluation, and answer those questions which arise concerning engineering aspects of the process.

Heating high-pyrite coals in aqueous slurry with compressed air at 1000-1200 psi and at 150°-160°C decreases pyritic sulfur to near the lower limit of detection by standard analytical procedure. Some results of 1 hr batch experiments are shown in Table I. The sulfur which is removed is converted primarily to aqueous sulfuric acid, with small amounts of ferric and ferrous sulfates in solution. At initial pressures of 800 psi and above and at temperatures above 160°C, 80% or more of the pyritic sulfur in the Minshall Seam (Indiana) coal is removed in 5 min at temperature. Similar pyritic sulfur removal was observed in a series of 10 min experiments at 1000 psi in the continuous unit.

TABLE I. Pyrite Removal from Representative Coals by Oxydesulfurization

Seam	State	Temp (°C)	Pyritic Sulfur (wt %) Untreated	Treated
Illinois No. 5	Illinois	150	0.9	0.1
Minshall	Indiana	150	4.2	0.2
Lovilia No. 4	Iowa	150	4.0	0.3
Pittsburgh	Ohio	160	2.8	0.2
Lower Freeport	Pennsylvania	160	2.4	0.1
Brookville	Pennsylvania	180	3.1	0.1

At 200 psi, the reaction is much slower, requiring several hours to achieve even 60% pyritic sulfur removal. For some coals, at least, the desulfurization is almost as rapid at 500 psi as at 1000 psi. The oxidation of pyritic sulfur is temperature dependent, but at the conditions of our experiments, reaction is sufficiently fast that above 150°C little improvement is noted. In a few cases, where a coal appears to have some residual pyrite which is not oxidized readily at 150°C, it may be removed at 180°C.

As the temperature at which the oxidation is conducted is increased above 150°C, an increasing amount of organic sulfur is removed from the coal. Although the percentage of organic sulfur removed parallels the temperature rise, so does the amount of coal which is oxidized. To prevent excessive loss of coal, a practical limit of 200°C has been chosen for carrying out the reaction on most coals. Removal of organic sulfur from a series of coals, shown in Table II, varies from 20 to over 40%. Further reduction of organic sulfur content is probably possible with some of these coals without sacrificing coal recoverability.

An upper limit on organic sulfur removal appears to be between 40 and 50%, and varies from coal to coal. We believe this is caused by the functionality of the organic sulfur, and gives some rough measure of oxidation resistant or DBT type of sulfur. Obviously, that sulfur which is removed by oxydesulfurization must be in some other structure which is readily oxidized, such as thiol, sulfide, and/or disulfide. A similar amount of organic sulfur is removed from coal when it is heated at 300°C with aqueous alkali, a reagent which does not attack DBT (5,6).

TABLE II. Organic Sulfur Removal from Representative Coals by Oxydesulfurization

Seam	State	Temp (°C)	Organic Sulfur (wt %)	
			Untreated	Treated
Bevier	Kansas	150	2.0	1.6
Mammoth[a]	Montana	150	0.5	0.4
Wyoming No. 9[a]	Wyoming	150	1.1	0.8
Pittsburgh	Ohio	180	1.5	0.8
Lower Freeport	Pennsylvania	180	1.0	0.8
Illinois No. 6	Illinois	200	2.3	1.3
Minshall	Indiana	200	1.5	1.2

a. Subbituminous.

Even at 150°-160°C many coals, including some with rather high sulfur contents, can be dramatically desulfurized as shown in Table III.

TABLE III. Oxydesulfurization of Representative Coals

Seam	State	Temp (°C)	Total Sulfur (wt %) Untreated	Total Sulfur (wt %) Treated	Sulfur (1b/10[6] Btu) Untreated	Sulfur (1b/10[6] Btu) Treated
Minshall	Indiana	150	5.7	2.0	4.99	1.81
Illinois No. 5	Illinois	150	3.3	2.0	2.64	1.75
Lovilia No. 4	Iowa	150	5.9	1.4	5.38	1.42
Mammoth[a]	Montana	150	1.1	0.6	0.91	0.52
Pittsburgh	Pennsylvania	150	1.3	0.8	0.92	0.60
Wyoming No. 9[a]	Wyoming	150	1.8	0.9	1.41	0.78
Pittsburgh	Ohio	160	3.0	1.4	2.34	1.15
Upper Freeport	Pennsylvania	160	2.1	0.9	1.89	0.80

a. Subbituminous.

The reaction conditions which we have found to be suitable for oxydesulfurization are: temperature between 150° and 220°C, operating pressure between 220 and 1500 psi, and residence time of 1 hr or less. Most of our experiments have been carried out below 220°C and at approximately 1000 psi. Recoveries of fuel values are excellent, generally 90% or better. The only by-product of the reaction is dilute H_2SO_4 containing iron sulfates. This can be recycled with no observable effect on desulfurization for at least five cycles. When the H_2SO_4 becomes too concentrated for further use, it can be converted to a commercial grade of sulfuric acid if a suitable, economic market exists, or it can be disposed of by limestone neutralization as a readily filterable $CaSO_4$.

The process, outlined in Figure 1, needs no novel technology to produce coal having over 95% of its pyritic sulfur and as much as 40% of its organic sulfur removed. Other than the coal, air, and water, the only other material needed for the process is the limestone used to neutralize the H_2SO_4. No sludge is formed, much of the water can be recycled, and the only waste product is solid $CaSO_4$. Similar oxidative desulfurizations using air and water (7) and oxygen and water (8) have been reported.

A preliminary estimate for this process indicates a cost of $8.00–$10.00 per ton. At this price, the process would still be considerably less expensive than coal conversion to gas or liquid fuel. Assuming removal of 95% pyritic sulfur and 40% organic sulfur, an estimated 40% of the coal mined in the eastern U.S. could be made environmentally acceptable as boiler fuel according to EPA standards for new installations. And the sulfur content of the remainder of the eastern coal could be drastically reduced, making it environmentally acceptable for existing boilers.

Conclusions

Treatment with compressed air and water at 150°-200°C represents a practical method to desulfurize to acceptable levels a sizable percentage of the available coal in the eastern United States at a cost in money and fuel value which is less than coal conversion and to an extent which is greater than can be achieved by physical depyriting methods.

Abstract

Both pyritic and organic sulfur in coal can be removed by a variety of oxidation techniques, including treatment with NO_x, peroxygen compounds, air in the presence of specific organic media, or air and water at elevated temperature and pressure. The most promising method involves contacting an aqueous slurry of coal with air at pressures up to 1000 psi

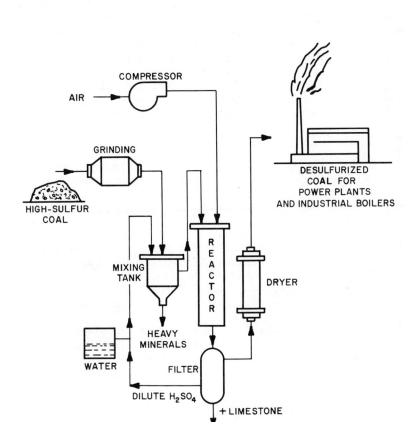

Figure 1. Air–water oxydesulfurization process

and temperatures of 140°-200°C. Coals from different geo-
graphic areas of the U.S. have been treated with air and water
in this way, resulting in removal of more than 90% of pyritic
sulfur and up to 40% of organic sulfur as sulfuric acid, which
is separated from the desulfurized coal by filtration. Fuel
value losses are usually less than 10%. Costs for processing
coal by this procedure will be somewhere between the less
efficient, less thorough and less costly physical coal cleaning
methods and the more thorough but much more costly coal
conversion techniques. Oxidative desulfurization potentially
can upgrade up to 40% of the bituminous coal in the U.S. to
environmentally acceptable boiler fuel and can bring most of
the rest of the bituminous coals at least close to acceptability
with relatively little loss in total fuel value.

Literature Cited

1. Reggel, L., Raymond, R., Wender, I., Blaustein, B. D.,
   Am. Chem. Soc., Div. Fuel Chem. Preprints, (1972) 17 (2),
   44.
2. Wallace, T. J., Heimlich, B. N., Tetrahedron (1968), 24,
   1311.
3. Reed, E., "Organic Chemistry of Bivalent Sulfur," Vol. II,
   p. 64, Chemical Publishing Co., Inc., New York, 1968.
4. LaCount, R. B., Friedman, Sidney, J. Org. Chem (1977), 42,
   2751.
5. Worthy, W., Chem. Eng. News, (July 7, 1975), pp. 24-25.
6. Friedman, Sidney, Warzinski, R. P., Trans. ASME, J. Eng.
   Power, (1977) 99A, 361.
7. Thomas, J., Warshaw, A., U.S. Pat. 3,824,084 (Assigned to
   Chemical Construction Corp.) July 16, 1974.
8. Agarwal, J. C., Giberti, R. A., Irminger, P. F., Petrovic,
   L. F., Sareen, S. S., Min. Congr. J., (1975) 61 (3), 40.

Sulfur Removal from Coals: Ammonia/Oxygen System

S. S. SAREEN[1]

Ledgemont Laboratory, Kennecott Copper Corp., 128 Spring Street, Lexington, MA 02172

The emergence of chemical desulfurization of coal as a viable alternative to stack gas scrubbing (1) has prompted researchers in this field to consider a variety of chemical systems which remove pyritic sulfur or both the pyritic and part of the organic sulfur (2,3,4,5,6). A review of the more prominent chemical desulfurization schemes is presented in a recent article (6). Because chemical desulfurization is cost competitive with stack gas scrubbing (1), utilities are beginning to show an interest in the development of this technology which could provide them with a source of clean fuel to meet the rigorous EPA standards for sulfur emissions.

This chapter discusses the sulfur removal from coals using an ammonia/oxygen/water system which removes almost all of the pyritic sulfur and up to 25% of the organic sulfur in about 2 hr. Because organic sulfur removal necessarily implies coal carbon losses, a balance must be struck between the amount of organic sulfur removed and the thermal losses that can be economically tolerated from the coals being cleaned.

Although no effort has been made to optimize the system reported here, the results of Btu loss, oxygen consumption, retention time, etc., are fairly consistent with the oxygen/water system for pyrite removal from coals (2). The carbon losses, as might be expected, are somewhat higher. Furthermore, the data presented here can be used to construct an optimization scheme for future development work.

Process Description

In this desulfurization scheme, run-of-mine coal is treated in a conventional preparation plant, where the coal is crushed and washed to remove rock and clay material. The coarsely crushed coal is then fed into closed-circuit wet ball mills where it is further ground to -100 mesh. The ground slurry is pumped into

[1]TRW Energy Systems, 7600 Colshire Drive, McLean, Va. 22101

oxygen-sparged leach reactors which operate at about 130^0C and
300 psi oxygen pressure. All of the pyritic sulfur and up to 25%
of the organic sulfur is removed in about 2 hr. The desulfurized
slurry now goes through a solid/liquid separation operation where
the coal and liquid are separated. Because of the formation of
sulfates and the absorption of some of the CO_2 (from coal oxidation) into the ammonia solution, this mixed sulfate/carbonate
stream must be regenerated to recycle the ammonia back into the
process. The ammonia regeneration may be accomplished by calci-
ning and/or steam stripping. Sulfur removal from coals as a
function of ammonia concentration and retention time is discussed
below.

Experimental Conditions

All of the experiments reported here were carried out in a
batch mode in high-pressure stirred autoclaves. The autoclaves
were equipped with baffles, and the speed of agitation was controlled with the help of a tachometer and was verified at fre-
quent intervals with a stroboscope. The system was heated with a
jacketed electric heater exterior to the chamber, and the temperature was controlled to within a couple of degrees with a temperature controller. The reaction mixture was cooled rapidly at the
end of the experiment with a cooling system fitted inside the
high-pressure reactors. The mode of operation was as follows:
120 gms (dry basis) of Illinois No. 6 coal was slurried in ammoniacal solutions to give a solids pulp density of 20 w/o. The
autoclaves were sealed and the air purged with inert gas. This
insured no reaction with the coal sulfur during the heatup period.
The heater and stirrers were turned on, and the temperature was
allowed to stabilize at 130^0C. The system was then pressurized
to yield an oxygen partial pressure of 300 psi, and the reaction
was permitted to proceed. The oxygen was aspirated from the vapor space of the autoclave, through the hollow agitator shaft, to
the impeller. The rotating impeller action resulted in the gas
being finely dispersed into the slurry. The vapor space of the
reactor was connected to a gas chromatograph for measurement of
the gas phase for products of reaction. Except for small quantities of gas removed for analytical purposes, there was no gas
bled from the system or any oxygen supplied during an experimental
run.
Minus 100 mesh coal was used, and the degree of agitation was
fixed to maintain the system in a kinetically controlled re-
gime (2). Ammonia concentrations between 0.5M and 5M were studied. An analysis of the starting Illinois No. 6 coal is listed
in Table I.

Table I. Starting Coal Analysis

Total sulfur 4.99%	ash 19.27%
Pyritic sulfur 2.06%	vol. matter 35.6%
Sulfate sulfur 0.65%	fixed carbon 45.13%
Organic sulfur 2.28%	BTU (maf) 13477

Experimental Results

The chemical reaction for the oxidation of pyrite in an ammoniacal system is given by Equation 1:

$$FeS_2 + 4NH_3 + 7/2\ H_2O + 15/4\ O_2 \rightarrow 4NH_4^+ + 2SO_4^= + Fe(OH)_3 \quad (1)$$

where all of the sulfide sulfur is oxidized to soluble sulfates. Care was taken to insure that the NH_3/FeS_2 molar ratio for the experimental study was always in excess of the 4 required stoichiometrically: a range between 6.5 and 65 was considered.

The effect of retention time and ammonia concentration on sulfur removal from Illinois No. 6 coals is graphically displayed in Figure 1. Approximately 90% of the pyritic sulfur can be removed, and there appears to be no apparent effect of NH_3 concentration on pyrite removal. There seems to be a definite trend, however, in the organic sulfur removal as a function of NH_3 concentration. Measurements of total change in sulfur content of the coal, expressed as lbs. SO_2/MMBtu, shows a 50% change between the starting coal and the desulfurized coal. This compares against a 25% change after desulfurization of Illinois No. 6 coals when using the O_2/H_2O system (2) where only pyritic sulfur is removed. The desulfurized coals, after an NH_3/O_2 treatment, also show no residual sulfate sulfur.

An important consideration in any chemical desulfurization process in which the coal sulfur is oxidixed is the oxidant consumption. For this process the oxygen consumption to oxidize the coal sulfur species and the coal itself may be listed as follows:

(1) Reaction with pyrite

(2) Oxidation of organic sulfur

(3) Oxygen uptake by the coal

(4) Oxidation of coal to form CO and CO_2 in the gaseous phase

(5) Formation of carbonates in solution

The stoichiometric oxygen consumption for the pyrite reaction, given by Equation 1, calculates to be 1.0 lb. O_2/lb. FeS_2.

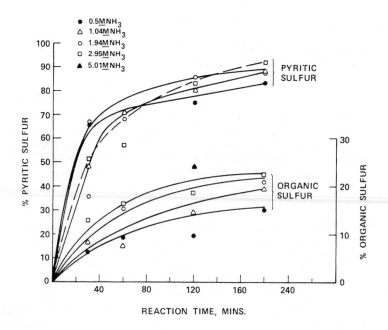

Figure 1.   Sulfur removal as a function of ammonia concentration and time

In the oxidation of pyrite with oxygen, it is inescapable that oxygen will also react with the coal carbon. This oxidation of the coal usually results in the formation of CO and CO_2, together with soluble coal acids. There is a greater propensity for the formation of coal acids in basic systems than in acid systems. Furthermore, there is some pickup of oxygen by the coal to form an intermediate oxygen-coal complex.

The gases, as analyzed in the vapor space of the autoclaves, using a gas chromatograph, show that CO formation is negligible and that CO_2 is the major product of reaction. Some of the CO_2 formed from carbon oxidation tends to dissolve in the ammoniacal liquor and report as carbonates in solution. To determine accurately the exact oxygen consumption for coal oxidation, solution analyses were conducted to measure this amount of carbonate formed. These analyses showed that the amount of carbonate measured in solution increased with increasing ammonia concentration. The total oxygen consumed by coal oxidation to form both CO_2 in the vapor space and carbonate in solution, as a function of retention time and ammonia concentration, is shown in Figure 2. These results show that coal oxidation is fairly insensitive to the ammonia concentration.

Because of the formation of an oxygen-coal complex, there is some oxygen tied up with the coal. This consumption is graphically displayed in Figure 3, which again shows no dependence on ammonia concentration. These oxygen values reflect the total oxygen content of the coal, for both the starting coals and after depyritization, as measured by neutron activation techniques. A minimum oxygen consumption for this process is tabulated in Table II.

Table II. Minimum Oxygen Consumption

	LBS. O_2/LB. COAL
O_2 for pyrite reaction	0.0375[a]
O_2 uptake by coal	0.0340[b]
O_2 for CO_2 + $CO_3^=$	0.0350[b]
	0.1065

[a] Based on 2% pyritic sulfur coal.

[b] After 2 hr. of sulfur removal.

This table does not take into account any oxygen consumption for organic sulfur oxidation, which is difficult to measure. For a plant processing 8000 tons/day of 2% pyritic sulfur coal (with a pyritic sulfur/organic sulfur ratio of 1), the oxygen demand based on Table II calculates to be 850 tons/day. With a 20% contingency

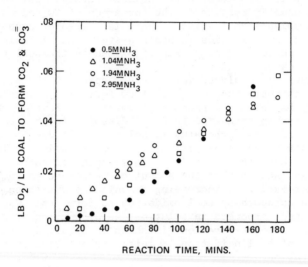

Figure 2.   *Oxygen consumption for coal oxidation*

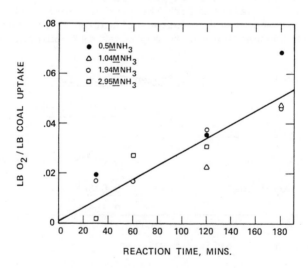

Figure 3.   *Oxygen uptake by coal*

factor to allow for organic sulfur oxidation, a 1000 tons/day oxygen plant would be needed. This total oxygen duty is the same as that required for coal desulfurization using the O_2/H_2O system (2).

An important consideration for the viability of any desulfurization process is the overall thermal efficiency of the system. The Btu loss, on a moisture-ash-free basis, as a function of reaction time and ammonia concentration is presented in Figure 4. This graph shows that between 8% and 13% loss may be expected after 2 hr. of sulfur removal. The pairing of the Btu loss as a function of ammonia concentration shown in Figure 4 is not immediately obvious. The loss of carbon during the desulfurization process is graphically displayed in Figure 5. In all cases the carbon loss is much greater than can be accounted for by CO_2 and carbonate formation. This suggests that the difference reports in solution as coal acids. The formation of these acids is not surprising since the reaction of alkalis with coal to form humic acids is well known.

A comparison between the key parameters of the H_2O/O_2 systems (only pyritic sulfur removal) and the NH_3/O_2 system (pyritic sulfur and some organic sulfur removal) is presented in Table III.

Table III. Comparison between H_2O/O_2 and NH_3/O_2 Systems[a]

	H_2O/O_2	NH_3/O_2
Sulfur removal		
pyritic	90^+%	90^+%
organic	0 %	25 %
Oxygen consumption (lbs. O_2/lb. coal)		
O_2 for pyrite reaction	0.035	0.0375
O_2 uptake by coal	0.054	0.0340
O_2 to form CO_2	0.032	0.0350
Btu loss (MAF)	8 %	13 %
Carbon loss (MAF)	6.5%	10%

[a]Illinois No. 6 coals, 20% slurry density, 2% pyritic sulfur.

The higher Btu and carbon losses for the NH_3/O_2 system are to be expected because organic sulfur removal necessarily implies some coal (and therefore Btu) losses.

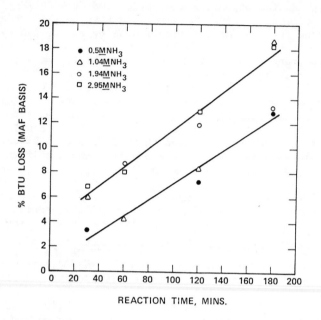

Figure 4. Btu loss from coals

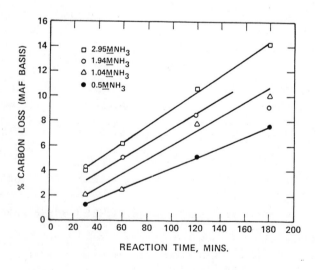

Figure 5. Carbon loss from coals

Summary and Conclusions

1. Increasing reaction time and ammonia concentration improved the extent of organic sulfur removal. For example, after 2 hr. using a $5\underline{M}$ NH_3 solution, 25% of the organic sulfur can be removed.

2. Changing the NH_3 concentration had no apparent affect on pyritic sulfur removal.

3. The oxygen consumed to oxidize the coal carbon (reporting as CO_2 in the gas phase and carbonate in solution) is fairly insensitive to the NH_3 concentration.

4. Ammonia concentration has no apparent affect on oxygen uptake by the coal.

5. Both the ammonia concentration and reaction time have an effect on the Btu and carbon values of coal. Increasing either one decreases the Btu and carbon value. As much as a 13% Btu and 10% carbon loss may be realized with using a $3\underline{M}$ NH_3 solution and reacting the coal for 2 hr. The large carbon losses are from the formation of coal acids.

Literature Cited

1. Agarwal, J. C., Giberti, R. A., Irminger, P. F., Petrovic, L. J., Sareen, S. S., <u>Min</u>. <u>Congr</u>. <u>J</u>., (1975), <u>61</u> (3), 40–43.

2. Sareen, S. S., Giberti, R. A., Irminger, P. F., Petrovic, L. J., <u>AIChE Symp. Ser</u>., (1977), <u>73</u> (165), 183–189.

3. Lorenzi, L. Jr., Land, J. S., VanNice, L. J., Koutsoukas, E. P., Meyers, R. A., <u>Coal</u> <u>Age</u>, (Nov. 1972), <u>77</u> (11), 70–76.

4. Diaz, A. F., Guth, E. D., U.S. Patent No. <u>3,909,211</u> (1975).

5. Smith, E. G., Am. Chem. Soc., Div. of Fuel Chemistry, 169th National Meeting, Philadelphia, Pa., April 6–11, 1975.

6. "Coal Cleaning Readies for Wider Sulfur-Removal Role," <u>Chem</u>. <u>Eng</u>., (March 1, 1976), <u>83</u> (6), 70–74.

15

Desulfurizing Coal with Alkaline Solutions Containing Dissolved Oxygen

C. Y. TAI,* G. V. GRAVES, and T. D. WHEELOCK

Iowa State University, Department of Chemical Engineering and Nuclear Engineering, Energy and Mineral Resources Research Institute, Ames, IA 50011

The extraction of pyritic sulfur from coal by leaching the comminuted material with hot aqueous solutions containing dissolved oxygen has been demonstrated in numerous laboratory experiments (1-6). Although acidic solutions have generally been used for such experiments, basic solutions appear to offer several important advantages. Thus Majima and Peters (7) showed that the rate of extraction of sulfur from relatively pure pyrite is much greater in basic solutions containing dissolved oxygen than in neutral solutions. Moreover it has been shown recently that basic solutions containing ammonium hydroxide and oxygen can extract a significant portion of the organic sulfur as well as most of the inorganic sulfur from coal at relatively moderate temperatures (e.g., 130°C) (4,5) whereas higher temperatures (150°-200°C) seem to be required with acidic solutions to remove organic sulfur (6). Furthermore some types of basic solutions are much less corrosive towards the common materials of construction than acidic solutions.

Although the chemical kinetics and mechanism of pyrite reaction with an oxygen-bearing caustic solution have been studied, they are not completely defined. Stenhouse and Armstrong (8) found sulfate ions and iron oxides to be the final products of reaction. Both Fe_2O_3 and Fe_3O_4 were identified in the residue. It appeared to these investigators that the iron oxides formed a stable layer around the unreacted pyrite. As a result of a later investigation, Burkin and Edward (9) concluded that the final oxidation product of iron is maghemite (γFe_2O_3) which is formed through a series of topotactic transformations. These investigators observed that the rate of attack was more rapid along grain boundaries and cracks in impure and imperfect pyrite crystals, with the result that these particles were leached more rapidly than particles containing few imperfections and impurities,

---

*Present address: Chung-Shen Institute of Science and Technology, Taiwan, R.O.C.

even though the insoluble oxidation product adhered to the imperfect crystals but not to the perfect ones. As a result of the
intergranular attack by the leach solution, the smaller crystals
within particles composed of different-sized crystals were desulfurized before the larger crystals. The reaction rate increased
with temperature according to the Arrhenius equation with an
activation energy of 9 kcal/mol. Also the reaction rate was con-
tolled by the oxygen concentration at the mineral-solution interface so the rate increased with the oxygen partial pressure. In
addition, the rate increased with increasing concentration of
sodium hydroxide up to 2 wt %. Higher concentrations reduced the
rate slightly.

In a solution of some base such as sodium hydroxide, the
stoichiometry of the reaction of pyrite with oxygen and the subsequent neutralization of the acid produced can probably be represented by the following equations:

$$FeS_2 + \frac{15}{4} O_2 + 2 H_2O = \frac{1}{2} Fe_2O_3 + 2 H_2SO_4 \tag{1}$$

$$2 H_2SO_4 + 4 NaOH = 2 Na_2SO_4 + 4 H_2O \tag{2}$$

Although these equations suggest that the main purpose of the
alkali is to neutralize the acid and to drive the first reaction
to completion, the actual reaction mechanism is probably more
complex with the alkali playing a more subtle role as well as the
obvious one.

An investigation by McKay and Halpern (10) of the reaction
of pyrite with oxygen in acidic solutions showed that the product
distribution, and therefore the reaction mechanism, is quite different from that noted above for basic solutions. Thus pyrite
was converted to soluble ferrous and ferric sulfate, sulfuric
acid, and elemental sulfur when leached with oxygen-bearing acidic
solutions at 100°-130°C. The product distribution depended on
temperature and acidity, with a lower temperature and higher acidity favoring the production of elemental sulfur and a higher temperature and lower acidity favoring the production of sulfuric
acid and soluble sulfates. Also at moderate pH (but not at low
pH), the ferric hydroxide hydrolyzed, and ferric oxide precipitated. The Arrhenius activation energy was determined to be 13
kcal/mol, which, being significantly larger than that noted above
for the reaction of pyrite in a caustic solution, indicates a
greater energy barrier for the reaction in an acidic medium.

The present investigation was undertaken to evaluate the
technical feasibility of extracting sulfur from coal with hot
oxygen-bearing solutions. Since it was soon confirmed that alkaline or basic solutions resulted in a much greater rate of extraction than acidic solutions, considerable emphasis was placed
on determining the effectiveness of different alkalis and alkali
concentrations. Several high-sulfur bituminous coals, as well as

pyrite which had been isolated from one of these coals, were leached in a small stirred reactor under a variety of experimental conditions. In addition the over-all effectiveness of chemical leaching in combination with physical beneficiation was investigated.

Experimental

Leaching experiments were carried out in a 1-L, stainless steel autoclave equipped with a removable nickel liner and an agitator consisting of two pitched blade turbine impellers (7.3 cm diameter) mounted on a shaft driven by a compressed air motor. The autoclave was also equipped with a pressure gauge, temperature indicator, and electric heating jacket.

A suspension of coal or pyrite particles in an alkaline solution of a specific concentration was placed in the reactor which was then sealed and heated to a specific operating temperature. After the system attained this temperature, oxygen was introduced from a high-pressure cylinder equipped with a combination pressure-reducing valve and regulator. During the succeeding operation, the system pressure was kept constant by supplying oxygen on demand. Also a small amount of gas was bled continuously from the reactor to avoid any build-up of gaseous reaction products in the system. Reaction conditions were kept constant for a specific time while the material was leached. At the end of the reaction period, the flow of oxygen into the system was stopped, and the system was cooled. Throughout the operation the agitator was kept running at a constant speed which was usually in the range of 250-350 rpm. After a run the leached solids were recovered, dried, and weighed. Coal samples were analyzed by the standard ASTM methods of analysis (11). When pyrite alone was leached, the amount of sulfur extracted was determined by measuring the sulfate content of the spent leachant using the sulfate titration method of Fritz and Yamamura (12).

Coal samples from the Big Ben, ICO, and ISU Demonstration Mine No. 1, all located in Southeastern Iowa, were leached. The rank of the coal from these mines is high volatile C bituminous coal; typical compositions are shown in Table I. Since the coal from these mines is very heterogeneous, various batches used for specific sets of experiments were re-analyzed and the composition reported along with other experimental results.

A handpicked sample of iron pyrites was obtained from the coal deposit at ISU Demonstration Mine No. 1. This material was very impure and seemed to contain about 74% FeS_2. Nevertheless, it was ground and screened into various sized fractions which were used subsequently for various leaching experiments. The sulfur content of the different size fractions is shown in Table II. In addition to pyritic sulfur, the material contained some sulfur in the form of sulfates and some in a form which was neither pyritic nor sulfate.

Table I. Proximate Analysis of Iowa Coals

Source of Coal	Proximate Analysis (wt %)			
	Volatile Matter	Fixed Carbon	Moisture	Ash
Big Ben mine	37.5	48.8	1.2	12.5
ICO mine	44.0	42.0	4.5	9.5
ISU mine	38.1	44.2	2.2	15.5

Table II. Composition of Iron Pyrites

Size Fraction (U.S. Series Mesh)	Composition (wt %)				
	Iron	Pyritic Sulfur	Sulfate Sulfur	Total Sulfur	Other
−60/+80				41.7	
−120/+140	42.5	39.2	0.5	40.9	16.6
−200/+230		36.4	0.7	40.3	
−230/+270		36.3	0.7	39.4	

Pyrite Leaching Experiments

A series of leaching experiments was carried out to determine the effect of various system parameters on the extraction of sulfur from impure pyrite. For each experiment a suspension consisting of 2.0 g of pyrite particles and 500 ml of an aqueous solution was treated in the stirred autoclave at 150°C with oxygen dissolved under a partial pressure of 3.27 atm. Since the vapor pressure of water at this temperature is 4.56 atm, the total system pressure was 7.8 atm. Unless indicated otherwise, the leaching operation under oxygen pressure at the specified temperature was conducted for 1 hr. The conversion of all forms of sulfur in the particles to soluble sulfate was determined by analyzing the spent leachant.

The results presented in Figure 1 indicate that among the various alkalis tested, sodium carbonate was the most effective. Essentially all of the sulfur was extracted from −200/+230 mesh

pyrite particles in 1.0 hr by an oxygenated solution containing
1.5 wt % sodium carbonate. The best results obtained with either
sodium hydroxide or sodium phosphate were slightly less favorable
and with ammonium carbonate much less favorable. The ammonium
carbonate solutions were noticeably less basic than the other
solutions and this factor probably accounted for the poorer results obtained with them. However, the ammonium carbonate solutions, as well as the other alkaline solutions, all gave better
results than water and oxygen alone. Thus only 27% of the sulfur
was extracted from the pyrite during a run made without any
alkali.

The effectiveness of all of the alkalis increased with initial concentration up to some optimum value and then decreased.
For sodium hydroxide the optimum initial concentration was about
1 wt %, for sodium carbonate 1.5 wt %, for sodium phosphate 3 wt
%, and for ammonium carbonate 4 wt %. The sharp increase in conversion for initial concentrations at the low end of the concentration scale seems to be related to the amount of alkali required to neutralize all of the acid produced in oxidizing the
pyrite. Thus if it is assumed that all of the sulfur in the
solids was converted into sulfuric acid during the leaching
operation, it would have required an initial sodium hydroxide
concentration of 0.40 wt % to neutralize all of the acid produced. Interestingly enough, the data presented in Figure 1 show
that the conversion fell off sharply when less than 0.40 wt %
caustic was used.

The amount of sodium carbonate required to neutralize the
acid depends on the final products of neutralization. Two possi-
bilities are shown below.

$$H_2SO_4 + Na_2CO_3 = Na_2SO_4 + CO_2 + H_2O \tag{3}$$

$$H_2SO_4 + 2 Na_2CO_3 = Na_2SO_4 + 2 NaHCO_3 \tag{4}$$

The first reaction would have required an initial sodium carbonate concentration of 0.53 wt % to neutralize all of the sulfuric
acid whereas the second reaction would have required an initial
concentration of 1.1 wt %. The experimental results (Figure 1)
indicate that the conversion fell off sharply when less than about
1 wt % sodium carbonate was used.

In the case of sodium phosphate an initial concentration of
1.65 wt % would have been required to neutralize all of the possible sulfuric acid according to the following reaction:

$$H_2SO_4 + 2 Na_3PO_4 = Na_2SO_4 + 2 Na_2HPO_4 \tag{5}$$

Although the experimental results shown in Figure 1 indicate that
the conversion fell off sharply between 2 and 3 wt %, the sparsity
of the data leave considerable doubt as to the exact critical

concentration.

Why the conversion declined at higher alkali concentrations which exceeded the optimum levels can not be explained with certainty. A similar decline was observed by Burkin and Edwards (9) when pyrite was leached with caustic solutions, and they attributed the decline to the reduced solubility of oxygen in these solutions. Also similar behavior was observed in leaching both galena and molybdenite with caustic solutions containing dissolved oxygen (13,14). In the case of both minerals, a decline in reaction rate with increasing alkali concentration was attributed to the decreasing solubility of oxygen.

The results of leaching different sized fractions of pyrite with sodium carbonate solutions of different concentrations are presented in Figure 2. Clearly the effect of particle size on conversion was very pronounced with the conversion increasing in an almost exponential manner as the size was reduced (Figure 2). Particles smaller than 60 μm in diameter were completely desulfurized in 1.0 hr by solutions containing 2-5 wt % sodium carbonate for the conditions shown in Figure 2. On the other hand, particles larger than 210 μm in diameter were only about half desulfurized under these conditions. Figure 2 also shows the advantage of using an alkaline solution. The lower curve in this diagram represented the conversion obtained when no alkali was added to the system. Even with the smallest particles of pyrite, only 35% of the sulfur was extracted in the absence of an alkali.

Several runs were made using longer reaction times to see how the conversion would change with time (Figure 3). The results indicate that the reaction rate decreased as the batch leaching operation proceeded. The declining rate could have been caused by the decreasing availability of unreacted pyrite, increasing resistance to diffusion of reactants within particles, increasing concentration of some rate-inhibiting reaction product in the leach solution, or a combination of these factors.

Coal Leaching Experiments

A series of leaching experiments was carried out to determine the effects of various alkalis and alkali concentrations on the desulfurization of coal from the ISU Demonstration Mine No. 1. The raw coal was pulverized so that 90% was finer than 200 mesh. For each experiment 50 g of coal and 500 ml of water or alkaline solution were placed in the stirred autoclave. The chemical treatment was carried out for 2 hr at 150°C with oxygen dissolved under a partial pressure of 3.27 atm and using a total system pressure of 7.8 atm. The treated coal was recovered, dried, weighed, and analyzed to determine the effectiveness of the treatment.

The best overall results were obtained with sodium carbonate among the following alkalis which were tested in various concentrations: sodium carbonate, sodium hydroxide, sodium phosphate,

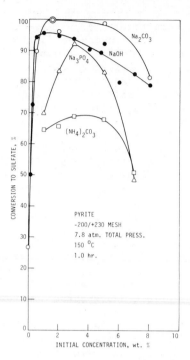

Figure 1.   The effect of different alkalis and alkali concentrations on the extraction of sulfur from pyrite

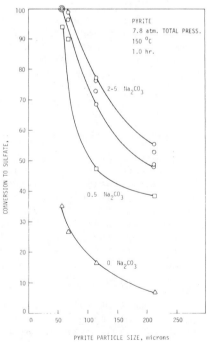

Figure 2.   The effect of particle size on the extraction of sulfur from pyrite

calcium carbonate, calcium hydroxide, and ammonium carbonate. The only alkali which approached sodium carbonate in effectiveness was sodium phosphate. However, it required a 7 wt % initial concentration of sodium phosphate to produce coal with a sulfur content comparable to that obtained with a 1.4 wt % initial concentration of sodium carbonate. Moreover, the Btu recovery was lower and the ash content of the product higher when sodium phosphate was used to obtain comparable desulfurization.

The initial concentration of sodium carbonate had a very pronounced effect on the extent of desulfurization and the resulting sulfur content of the treated coal (Table III, Figure 4). The best results were obtained with an initial concentration of 1.4 wt % and using either lower or higher concentrations resulted in less desulfurization. The treatment with the optimum concentration reduced the pyritic sulfur content of the coal by two-thirds. A similar effect of concentration was observed when sodium hydroxide was used with the optimum initial concentration being about 1 wt %. Hence, the effect of alkali concentration was like that experienced in leaching pyrite (Figure 1).

Under the mild leaching conditions used in this work, mainly inorganic sulfur was extracted from the coal. Essentially all of the sulfate form of sulfur was extracted and a good share of the pyritic sulfur. Unfortunately the results reported in Table III for organic sulfur are so erratic that it is not possible to discern whether organic sulfur was removed or not.

For the experiments involving sodium carbonate, the yield of dry coal was 93-94%, and the Btu recovery was 88-92%. These values could have been higher with the exercise of greater care in conducting the experiments. Since the heating value of the treated coal on a moisture and ash free (maf) basis was nearly the same as that of the untreated coal, it does not appear that the coal was significantly degraded by the treatment. Also the oxygen content of the coal, as determined by ultimate chemical analysis, did not appear to be significantly different from that of the untreated coal. These results contrast markedly with those reported by Sareen et al. (3,4) for treatments using much higher oxygen partial pressures which resulted in noticeable oxygen uptake by Illinois No. 6 coal.

As a result of the alkali treatment, the sodium and ash content of the coal increased measurably (Table III, Figure 5). The increase in ash content was directly proportional to the increase in sodium content and seemed caused almost entirely by the adsorption of sodium. Most of the increase in sodium or ash content took place at lower alkali concentrations, and more than 2 wt % sodium carbonate in the leachant generally produced little additional increase. In other words, the coal seemed to become saturated when the leach solution contained 2 wt % sodium carbonate or more. On the other hand, the coal retained very little sodium when the leach solution initially contained 0.8 wt % sodium carbonate or less because the solution became acidic by the

Table III. Results of Leaching ISU Coal with Different Concentrations of Alkali

Treatment	Na$_2$CO$_3$ Conc. (wt %)	Wt Yield (%)	Btu Recov. (%)	Heat Value (Btu/lb)		Ash (wt.%)	Na (wt %)	Sulfur Distrib. (lb/10^6 Btu)			
				as recvd.	maf			Pyritic	Sulfate	Organic	Total
none	--	--	--	11,240	13,660	15.5	0.05	3.22	0.60	1.81	5.63
wash	--	--	--	11,577	13,970	15.3	0.03	3.21	0.17	2.25	5.63
chemical	0	95	88	10,440	--	17.5	--	2.28	0.11	2.89	5.29
chemical	0.8	94	92	10,950	13,430	16.2	0.10	1.62	0.05	1.80	3.47
chemical	0.8	94	90	10,740	--	16.3	0.10	1.89	0.03	2.32	4.23
chemical	1.0	93	91	11,010	--	16.4	0.33	1.46	0.05	2.39	3.90
chemical	1.4	93	89	10,780	13,520	17.7	0.80	1.11	0.06	1.81	2.99
chemical	2.0	93	88	10,580	13,520	17.8	1.14	1.23	0.07	1.85	3.15
chemical	5.0	93	90	10,880	13,800	18.5	1.29	1.75	0.11	1.80	3.66
chemical	8.0	94	89	10,590	--	19.4	--	1.76	0.09	2.39	4.24
chemical	10.0	94	92	10,950	--	19.7	1.90	1.63	0.11	2.28	4.02
chemical	15.0	94	91	10,940	--	18.2	1.52	2.34	0.18	2.65	5.17

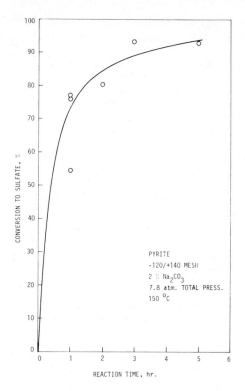

PYRITE

-120/+140 MESH

2 % Na_2CO_3

7.8 atm. TOTAL PRESS.

150 °C

Figure 3. The effect of reaction time on the extraction of sulfur from pyrite

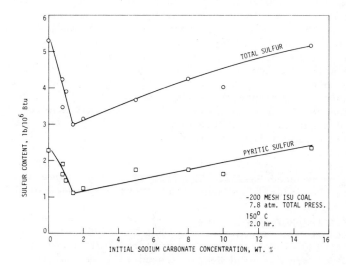

-200 MESH ISU COAL

7.8 atm. TOTAL PRESS.

150° C

2.0 hr.

Figure 4. Sulfur content of ISU coal after leaching

end of a run. Washing with water did not remove sodium from the
treated coal, but washing with dilute acid (0.1N hydrochloric
acid) did remove it.

Another series of experiments was conducted to see what
effect reducing the size of the coal would have on desulfurization
by chemical leaching. ISU coal was first pulverized to -35 mesh
(about 50% below 200 mesh) with a Mikro-Samplmill (Pulverizing
Machinery Division, American-Marietta Co.). The coal was then
wet-ground in a ceramic jar mill for different time periods. The
particle size distribution of the coal after grinding for 2, 4,
8, and 20 hr is shown in Figure 6. After grinding for 2 hr, 97%
of the coal was finer than 400 mesh (38 μm) size. The pulverized
or ground coal was then leached in the stirred autoclave for 2 hr
at 140°C with a 2.0 wt % solution of sodium carbonate containing oxygen dissolved under a partial pressure of 2.4 atm and with
a total system pressure of 5.8 atm. For each leaching run 25 g
of coal was mixed with 500 ml of solution. The results presented
in Table IV show that more sulfur was extracted from coal which
had been ground in the ball mill than from coal which had only
been pulverized with the Mikro-Samplmill. Thus the pyritic sulfur content of the ball-milled coal was reduced 87% whereas the
pyritic sulfur content of the pulverized coal was reduced only
71%. However, grinding the coal for more than 2 hr did not produce any additional benefit.

Combined Physical and Chemical Cleaning

A final set of experiments involved combined physical and
chemical cleaning of several Iowa coals. Each coal was crushed
to 6 mm x 0 size with a double roll crusher and then cleaned by
gravity separation in an organic liquid having a specific gravity
of 1.60. The cleaned float fraction was recovered, dried, and
pulverized with the Mikro-Samplmill. The pulverized coal was wet-
ground next in the ceramic jar mill for 20 hr and finally leached
in the stirred autoclave for 2.0 hr with a 2.0 wt % sodium carbonate solution. The effects of different leaching temperatures
and oxygen partial pressures were studied.

The results presented in Table V show that gravity separation
removed a substantial part of the pyritic sulfur from each coal
and that chemical leaching removed most of the remaining inorganic
sulfur, both pyritic and sulfate. Thus through the combined
treatment, the inorganic sulfur content of each of the three
coals was reduced 93-95%. The treated ICO coal came close to
meeting the federal new source performance standard of 0.6 lb
S/10^6 Btu.

For the physically cleaned and finely ground coal, it did
not seem to make any difference what leaching temperature in the
range of 120°-150°C or what oxygen partial pressure in the range
of 2.4-5.2 atm was used because essentially all of the inorganic
sulfur was removed even under the mildest treatment conditions.

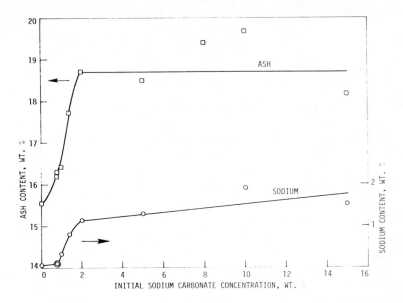

Figure 5. Ash and sodium content of ISU coal after leaching

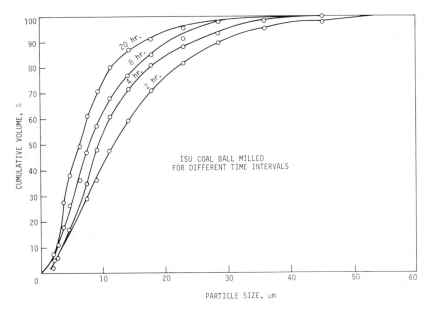

Figure 6. Cumulative volume of particles smaller than the indicated size

Table IV. Results of Treating Different Grinds of ISU Coal

Run No.	Treatment	Grind	H.V. (Btu/lb)	Ash (wt %)	Sulfur Distribution (lb/10⁶ Btu)			
					Pyritic	Sulfate	Organic	Total
3-108-5	none	pulverized	11,040	12.0	2.31	0.54	2.19	5.04
3-108-5	chemical	pulverized	11,360	15.8	0.67	0.05	2.32	3.05
3-108-7	chemical	b.mill (2 hr)	11,500	14.7	0.29	0.04	2.10	2.43
3-108-6	chemical	b.mill (4 hr)	11,070	16.6	0.30	0.05	2.17	2.51
3-108-8	chemical	b.mill (8 hr)	11,237	16.2	0.34	0.05	2.06	2.46

Table V. Results of Combined Physical and Chemical Treatment

Coal			Temp. (°C)	Pressure (atm.)		H.V. (Btu/lb)	Ash (wt %)	Sulfur Distrib. (lb/10^6 Btu)			
Source	Amt.(g)	Treatment		O_2	Total			Pyritic	Sulfate	Organic	Total
ISU	--	none	--	--	--	10,990	--	3.96	0.33	2.53	6.81
ISU	--	float/sink	--	--	--	12,390	--	1.38	0.23	2.41	4.02
ISU	40	chemical	120	2.4	4.4	11,650	12.2	0.13	0.06	2.34	2.53
ISU	40	chemical	135	2.7	5.8	11,490	12.1	0.18	0.07	2.32	2.57
ISU	20	chemical	150	2.4	6.9	10,770	24.0	0.21	0.04	2.04	2.29
ICO	--	none	--	--	--	12,040	--	1.66	0.25	0.61	2.52
ICO	--	float/sink	--	--	--	12,690	--	0.70	0.20	0.62	1.53
ICO	40	chemical	120	2.7	4.7	11,630	11.6	0.10	0.02	0.64	0.76
ICO	40	chemical	120	3.7	5.8	11,480	11.6	0.15	0.02	0.58	0.75
ICO	40	chemical	120	5.2	7.3	11,640	10.8	0.09	0.02	0.64	0.74
ICO	25	chemical	150	3.2	7.8	11,220	14.1	0.14	0.03	0.60	0.77
ICO	25	chemical	150	3.2	7.8	11,740	13.2	0.11	0.04	0.56	0.71
Big Ben	--	none	--	--	--	11,800	--	1.37	0.37	0.61	2.36
Big Ben	--	float/sink	--	--	--	12,790	--	0.42	0.28	0.81	1.47
Big Ben	25	chemical	132	2.9	5.8	10,950	10.6	0.09	0.03	0.89	1.00
Big Ben	25	chemical	150	3.2	7.8	11,680	10.0	0.16	0.04	0.81	1.01

Therefore, it should have been possible to obtain as good results even with milder treatment conditions or a shorter reaction time.

Conclusions

Dilute alkaline solutions containing oxygen dissolved under pressure were very effective at elevated temperatures for extracting the inorganic sulfur from coal. Of various alkalis tested, sodium carbonate gave the best results. This material also has the advantages of being readily available, low in cost, and relatively noncorrosive in aqueous solution towards steel and other common materials of construction. Although sufficient alkali should be used to neutralize all of the acid which is produced through oxidation of the coal sulfur, an excessive concentration of alkali seems to slow the rate of desulfurization. On the other head, the rate of desulfurization can be increased by reducing the size of the coal and/or pyrite particles. Leaching coal fines with dilute sodium carbonate solutions at temperatures up to 150°C and with oxygen partial pressures up to 5 atm for up to 2 hr does not seem to degrade high volatile bituminous coal. However, the coal does adsorb sodium from the leach solution which can be removed subsequently by washing with dilute acid. Chemical leaching can be combined advantageously with physical cleaning since the latter is more adept at removing coarser particles of pyrite while the former is more adept at removing the microscopic particles.

Acknowledgement

This work was sponsored by the Iowa Coal Project and conducted in the Energy and Mineral Resources Research Institute at Iowa State University.

Literature Cited

1. Nelson, H. W., Snow, R. D., Keyes, D. B., Ind. Eng. Chem. (1933) 25, 1355.
2. Agarwal, J. C., Giberti, R. A., Irminger, P. F., Petrovic, L. J., Sareen, S. S., Min. Congr. J. (1975) 61 (3), 40.
3. Sareen, S. S., Giberti, R. A., Irminger, P. F., Petrovic, L. J., Preprint 50c, 80th National Meeting AIChE, (Sep. 7-10, 1975), Boston.
4. Agarwal, J. C., Giberti, R. A., Petrovic, L. J., U.S. Patent No. 3,960,513, 1976.
5. Sareen, S. S., This volume, in press.
6. Friedman, S., LaCount, R. B., Warzinski, R. P., This volume, in press.
7. Majima, H., Peters, E., Trans. Metall. Soc. AIME (1966) 1409.
8. Stenhouse, J. E., Armstrong, W. M., Can. Min. Metall. Bull. (1952) 45, 49.

9. Burkin, A. R., Edwards, A. M., in "Mineral Processing," (A. Roberts, Ed.), pp. 159-169, Pergamon, Oxford, 1965. (Procedding of the Sixth International Congress held at Cannes, May 26-June 2, 1963.)

10. McKay, D. R., Halpern, J., Trans. Metall. Soc. AIME (1958) 301.

11. Book ASTM Stand., Part 26, "Gaseous Fuels; Coal and Coke; Atmospheric Analysis," Methods D2015-66, D2492-68, D3174-73, D3177-75, 1976.

12. Fritz, J. S., Yamamura, S. S., Anal. Chem. (1955) 27, 1461.

13. Anderson, J. E., Halpern, J., Samis, C. S., J. Met. (1953) 5, 554.

14. Dresher, W. H., Wadsworth, M. E., Fassell, W., Jr., J. Met. (1956) 8, 794.

16

Hydrothermal Coal Process

EDGEL P. STAMBAUGH

Battelle, Columbus Laboratories, 505 King Avenue, Columbus, OH 43201

Coal is the major source of energy for the U.S. and will continue to be so for many years. However, coal is dirty, containing high concentrations of contaminants such as sulfur, nitrogen, and mineral matter. These contaminants, if not removed from the coal before or during combustion, will find their way into the environment and thus constitute a serious health hazard.

An alternative to insure a healthier environment, as the consumption of coal as the major source of energy increases, is to remove these contaminants by chemical coal cleaning before combustion. One such method based on hydrothermal technology is the hydrothermal coal process in which certain coals can be chemically cleaned to produce solid fuels which meet Federal sulfur emission standards for new sources.

Process Description

The basic process, as shown schematically in Figure 1, comprises five major processing operations: coal preparation, hydrothermal (desulfurization) treatment, liquid/solid separation, fuel drying, and leachant regeneration. Coal preparation may entail a simple grinding operation to reduce the raw coal to the desired particle size of 70% -200 mesh or 100% -28 mesh. On the other hand, this operation may involve two operations—grinding of the coal to the desired particle size followed by physical beneficiation to remove a portion of the mineral matter including a portion of the pyritic sulfur.

Hydrothermal treatment entails basically three processing steps:

(1) The ground coal is mixed with an aqueous alkaline leachant, for example, an aqueous solution/slurry of sodium hydroxide and lime, to produce a raw coal slurry.

(2) This raw slurry is heated in an autoclave at about 250°-350°C (steam pressure of 600-2500 psig) to extract a significant portion of the sulfur and the mineral matter, depending on the leachant and processing conditions.

(3) The coal product slurry is cooled and pumped into a receiving tank.

The desulfurized coal is then separated from the spent leachant and washed with water. The final product is a solid fuel containing reduced concentrations of sulfur and, depending on the leaching conditions, mineral matter.

Drying the fuel to remove the residual moisture is optional. For some uses, it may be desirable to burn the fuel wet; in others, removal of a part or all of the residual moisture may be desirable. In any event, the coal can be dried in commercially available dryers.

The spent leachant containing the extracted sulfur as sodium sulfide (Na_2S) can be regenerated for recycle in several ways; one approach involves treatment of the spent leachant with carbon dioxide to remove the sulfur as H_2S which is subsequently converted to elemental sulfur by the Claus or Stretford process. The resulting liquor containing sodium carbonates is treated with lime to convert the carbonates back to sodium hydroxide which is recycled to the desulfurization segment after adjusting the concentration. The calcium carbonate is thermally decomposed to lime and carbon dioxide for recycle to the leachant regeneration segment.

Capability of Process

Sulfur Extraction. The process is capable of converting, on a laboratory (batch) and miniplant (1/4 ton per day) scale, a variety of major seam coals of different ranks and lignite to clean solid fuels having a total sulfur content equivalent to or less than 1.2 lb $SO_2/10^6$ Btu as shown in Table I. This is achieved by extracting greater than 90% of the pyritic sulfur from a large number of coals and extracting up to 50% of the organic sulfur from certain coals. Btu recovery as solid clean fuel is in the range of 90-95%, depending on the coal and processing conditions. The remainder, solubilized by the leachant, is recovered during leachant regeneration. This material could be used as process heat. Thus, a wide variety of major seam, high-sulfur bituminous coals and subbituminous coals in addition to lignite can be converted to environmentally acceptable solid fuels.

These coals can be used directly as a source of energy without further cleaning during the combustion process, assuming all of the sulfur is emitted to the atmosphere.

With alkali hydrothermally treated coals, all of the sulfur is not emitted to the atmosphere. During the desulfurization operation, the coal structure is opened up to give a product having a sponge-like morphology (see Figure 2). The porous structure allows the alkali to penetrate the coal particles and subsequently to react with the functional groups, for example, carboxylic acid groups, of the coal molecules. Also, the alkali

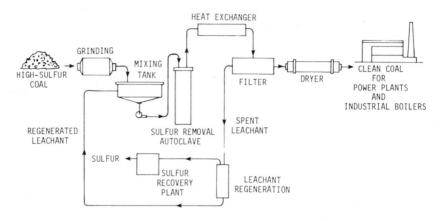

Figure 1.   Schematic of hydrothermal coal process

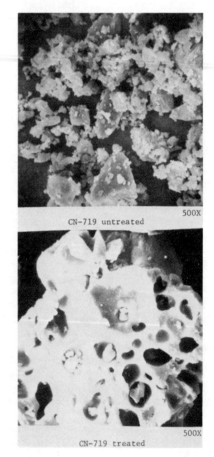

Figure 2.   Comparison of morphology of untreated (top) and HTT coal (bottom)

Table I.   Sulfur Emissions of Low-Sulfur Coals
from Hydrothermal Coal Process

Coal Source Seam	SO_2 Equivalent ($lb/10^6$ Btu)	
	Raw Coal	HTT Coal
Laboratory scale		
Lower Kittanning	2.2	0.9
Upper Freeport	2.4	0.9
Ohio 6	3.9	1.2
Pittsburgh 8	4.6	0.9
Pittsburgh	3.4	0.7
Lignite	1.5	1.2
Western	1.0	0.3
Prepilot plant		
Lower Kittanning	4.0	1.1
Upper Freeport	2.4	0.8

is physically deposited within and on the coal particles. Consequently, in addition to having a reduced sulfur content, the hydrothermally treated coal is impregnated with alkali which acts as a sulfur scavenger during the combustion process. In some recent combustion studies supported by EPA (1), 41.4-75.7% of the sulfur remaining in the treated coal was captured by the alkali in the treated coals, as shown in Table II. (This combustion study was conducted in a 1-lb/hr laboratory combustion facility.) Stack gases contained 120-290 ppm SO_2 equivalent to about 0.3-0.6 lb $SO_2/10^6$ Btu. Theoretically, assuming no sulfur capture, the stack gases would have contained 490-650 ppm SO_2, equivalent to about 1.0-1.5 lb $SO_2/10^6$ Btu. This added feature of the hydrothermal process greatly increases the number of coals which can be used as a source of clean energy, especially those coals containing relatively high concentrations of organic sulfur not subject to removal by mechanical cleaning.

Metal Extraction. Hydrothermal treatment results in the extraction of a number of the trace metals. Examples of these are shown in Table III. Therefore, trace metal emissions from the combustion of hydrothermally treated coal should be lower than those from the combustion of the corresponding raw coals.
    Further metal extraction is achieved by chemically deashing the hydrothermally treated coal with dilute acid, for example, sulfuric acid. As an illustration, the ash content of untreated coals from Ohio, Kentucky, West Virginia, and Pennsylvania ranged from 4.6-13.2%; ash content of chemically deashed hydrothermally treated coals ranged from 0.7-5.3%.

Improved Combustion Behavior. Certain general combustion characteristics of both the raw and treated coals, such as

Table II. Sulfur Capture by Alkali in HTT Coals[a]

	Martinka Raw Coal	Martinka Coal Caustic	Martinka Coal Mixed Leachant	Westland Raw Coal	Westland Coal Caustic	Westland Coal Mixed Leachant
Theoretical SO_2 (ppm)	1877	493	495	1610	580	651
SO_2 in stack gas (ppm)	1910	120	290	1115	250	220
Sulfur capture (%)	--	75.7	41.4	30.7	56.9	66.2

[a]Martinka coal represents the Lower Kittaning Seam; Westland coal represents the Pittsburgh Seam.

Table III.   Toxic Metals Extracted by Hydrothermal
Treatment of Coal

Metal	Concentration (ppm)	
	Raw Coal[a]	Treated Coal[a]
Arsenic	25	2
Beryllium	10	3
Boron	75	4
Lead	20	5
Thorium	3	0.5
Vanadium	40	2

[a] Average values from 3 Ohio coals.

ignition temperature and reactivity, were determined quantitatively from the derivative thermogravimetric (dTGA) and the
differential thermal (DTA) fuel analyses. The results of the
dTGA and DTA are summarized in Tables IV and V, respectively.

From these analyses, the combustion characteristics of these
coals in terms of ignition, reactivity, and possibly flammability
may have been improved by the hydrothermal treatment. For
example, the ignition temperature of Westland coal was reduced
from 426°-344°C (Table IV), a reduction of 82°C, by treating the
coal with sodium hydroxide and the mixed leachant systems. A
similar effect was noted by hydrothermal treatment of the Martinka
coal with these leachant systems.

This was expected in view of other hydrothermal work which
has been conducted at Battelle-Columbus. In this work, hydrothermal treatment of coals resulted in alteration and modification
of the coal structure to a more simplified structure. This is
evidenced by the fact that the liquid products from the pyrolysis
of HTT coals contained less asphaltenes than the liquid products
from the corresponding raw coals (2). The lower molecular weight
organic liquids from the HTT coals should have a lower ignition
temperature and a higher degree of flammability than the higher
molecular weight liquids from the raw coals.

The increased reactivity is reflected in Table V. For
example, treatment of the Martinka and Westland coals with the
mixed leachant system resulted in HTT coals which burned out at
a maximum of about 470°C whereas the raw coals burned out at
about 585°-600°C. A similar effect, but not to this degree, was
observed with the sodium hydroxide-treated coals.

While there may not be a direct correlation between
combustion and gasification, hydrothermal treatment of coal with
the mixed leachant system increases steam gasification and
hydrogasification rates by as much as 40-50 fold (3). This has
been attributed to alteration and modification of the coal
structure and to impregnation of the coal particle with a
catalyst, in this case calcium and/or sodium. This work has also

Table IV. Differential Thermal Analyses of Coal Samples in an Atmosphere of Air

	Westland Raw Coal Low Ash	Martinka Raw Coal	Westland Coal Sodium Treated	Martinka Coal Sodium Treated	Westland Coal Mixed Leachant	Martinka Coal Mixed Leachant
Starting exotherm (°C)	233	243	252	263	268	252
Ignition point (°C)	426	432	344	360	344	376
Secondary exotherm (°C)	-	-	488	508	494	493
End of exotherm (°C)	615	622	564	578	555	553

TABLE V. Thermogravimetic Analyses of Raw and HTT Coals[a]

	Westland Raw Coal	Martinka Raw Coal	Westland Coal NaOH Leachant	Martinka Coal NaOH Leachant	Westland Coal Mixed Leachant	Martinka Coal Mixed Leachant
Temperature range[b]	220-585	250-600	230-570	240-510	240-465	270-470
Maximum rate of weight loss (mg/min)	17.5	19.0	21.5	27.5	23.0	27.0
Temperature at maximum rate of weight loss (°C)	320	275	305	275	285	310

a TGA performed with Cahn Electrobalance at 15°C/min and air flow of 800 ml/min.
b Temperature range over which most of the sample is lost.

shown that the mixed leachant-treated coal is more reactive than
the sodium hydroxide-treated coal.

Summary

    Hydrothermal processing is a potential technology for
converting a variety of major seam, high-sulfur coals and lignite
to clean solid fuels having a total sulfur content equivalent to
or less than 1.2 lb $SO_2/10^6$ Btu. During the treatment, the coals
are impregnated with alkali which acts as a sulfur scavenger,
thereby reducing sulfur emissions still further during combustion.
The process is also effective in removing trace metals such as
beryllium, boron, vanadium, and arsenic; therefore, the gaseous
emissions from combustion of HTT coals should be less polluting
in terms of sulfur and trace metal emissions than from raw coals.
Hydrothermal treatment also appears to improve the combustion
characteristics of coal in terms of ignition, reactivity, and
possibly flammability.
    Thus, hydrothermal treatment converts at least certain high-
sulfur coals to environmentally acceptable solid fuels with
potentially improved combustion characteristics. These coals are
potential sources of energy for pulverized, stoker, and coal-
slurry combustion.

Acknowledgement

    Results as reported were supported in part by the Battelle
Energy Program and in part by the U.S. Environmental Protection
Agency, Research Triangle Park, NC, under Contract No. 68-02-2119.

Literature Cited

1.   Stambaugh, E. P., Levy, A., Giammar, R. D., Sekhar, K. C.,
     "Hydrothermal Coal Desulfurization with Combustion Results,"
     Proceedings of the Fourth National Conference (October 3-7,
     1976) 386-394.
2.   Stambaugh, E. P., Feldmann, H. F., Liu, K. T., Sekhar, K. C.,
     "Novel Concept for Improved Pyrolysis Feedstock Production,"
     173rd National Meeting American Chemical Society, New Orleans,
     Louisiana (March 21-25, 1977).
3.   Stambaugh E. P., Miller, J. F., Tam, S. S., Chauhan, S. P.,
     Feldmann, H. F., Carlton, H. E., Nack, H., Oxley, J. H.,
     "Battelle Hydrothermal Coal Process," 12th Air Pollution and
     Industrial Hygiene Conference on Air Quality Management in
     EPI, University of Austin, Austin, Texas (January 1976).

17

Coal Desulfurization by Low–Temperature Chlorinolysis

GEORGE C. HSU, JOHN J. KALVINSKAS, PARTHA S. GANGULI,
and GEORGE R. GAVALAS*

Jet Propulsion Laboratory, California Institute of Technology,
4800 Oak Grove Drive, Pasadena, CA 91103

Since most of the coals in this country, particularly the eastern and mid-western coals, have a high sulfur content (>2%), there is a need for an economical process of converting high-sulfur coals to clean fuel (<1.2 lbs of SO_2 emission per million Btu by EPA standards) so that coal can be used as a source of energy without causing serious air pollution.

Among the three principal methods for precombustion desulfurization of coal - physical depyriting, chemical desulfurization (1,2,3,4), and coal conversion to low-sulfur liquid and gaseous fuels, the potential of chemical methods looks promising in terms of both total sulfur removal and processing cost. The principal chemical methods for coal desulfurization involve treatment with either oxidizing agents (1,2,3) or basic media (4) at elevated temperature and pressure. For example, the method of oxidation (1) by contacting an aqueous slurry of coal with air at pressures up to 1000 psi and temperatures of 140° - $200^\circ C$ can remove more than 90% of the pyritic sulfur and up to 40% of the organic sulfur as sulfuric acid. Fuel value losses are usually less than 10%.

Most of the earlier studies (5,6,7,8,9) on chlorinolysis of coal were conducted to understand the chemistry of the process, to produce a non-caking fuel suitable for burning, and to investigate the possible production of chlorohydrocarbons from coal. However, there was a limited amount of work (10,11) on desulfurization by chlorination of coal in the gas phase at high temperature ($400^\circ C$) and elevated pressure. These studies showed that both organic and inorganic sulfur could be removed to a certain extent, but the loss of coal was more than 20% and because of high-temperature chlorination, satisfactory dechlorination afterwards was not achieved. A few studies (12) on the chlorination of coal in an aqueous media at $25^\circ C$ resulted in poor removal of sulfur.

* Professor, Chemical Engineering Department,
California Institute of Technology, Pasadena, California 91125

This chapter is an updated version of the presentation (13) given in 1976 at the ACS Centennial Meeting at San Francisco. It includes some recent experimental results showing the feasibility of removing sulfur, particularly organic sulfur, from high-sulfur coals by a simple method of low-temperature chlorinolysis followed by hydrolysis and dechlorination. The chemical feasibility of sulfur removal by chlorinolysis rather than the detailed engineering process is emphasized. The concept of this chlorination method is described, and experimental results and discussion of this method of desulfurization are presented for two bituminous coals.

On the basis of experimental data, the low-temperature chlorinolysis desulfurization method provides a number of basic advantages that include:

(1) A high degree of sulfur removal, particularly organic sulfur removal;

(2) Additional removal of trace metal concentrations contained in the original coal that reduces potentially hazardous emissions to the atmosphere;

(3) A relatively simple coal-processing operation mostly at low temperatures and atmospheric pressure which translates into a low-cost process;

(4) A beneficiation of the coal for use in direct-fired combustion as well as in gasification processes;

(5) Improved coal feedstocks for combustion and gasification operations since treated coal is non-caking and non-swelling.

Concept of Coal Desulfurization by Low-Temperature Chlorinolysis

Chlorination. The essential, although considerably simplified, desulfurization reactions involved during chlorinolysis of coal, as identified in sulfur and polymer chemistry literature and supported by Jet Propulsion Laboratory experimental results, may be illustrated as follows:

Because of high steric accessibility of bicovalent sulfur and the electron-releasing and electron-demanding nature of the sulfur atom, the carbon-sulfur (sulfide) and sulfur-sulfur (disulfide) bonds in coal will be highly reactive. The scission of carbon-sulfur and sulfur-sulfur bonds in organic compounds has been experimentally demonstrated in polymer and sulfur chemistry, (14)

C-S Bond Cleavage:

$$R-S-R' + Cl^+ - Cl^- \xrightleftharpoons{H^+} RSCl + R'Cl \qquad (1)$$

where R and R' represent hydrocarbon groups, and S stands for sulfur.

S-S Bond (electrophilic cleavage):

$$RS-SR' + Cl^+ - Cl^- \xrightleftharpoons{H^+} RSCl + R'SCl \qquad (2)$$

Conversion to Sulfates:

Sulfenyl chloride is oxidized to sulfonate or sulfate according to the following reactions:

$$RSCl \xrightarrow{Cl_2, H_2O} RSO_2Cl \xrightarrow{H_2O} RSO_3H + HCl \qquad (3a)$$

$$RSCl + 2Cl_2 + 3H_2O \longrightarrow RSO_3H + 5HCl \qquad (3b)$$

or

$$RSCl \xrightarrow{Cl_2, H_2O} RSO_2Cl \xrightarrow[Cl_2, H_2O]{\Delta} SO_4^= + RCl \qquad (4a)$$

$$RSCl + 3Cl_2 + 4H_2O \longrightarrow RCl + H_2SO_4 + 6HCl \qquad (4b)$$

Pyritic Sulfur:

Reactions (15) are summarized as follows:

$$FeS_2 + 2Cl_2 \longrightarrow FeCl_2 + S_2Cl_2 \qquad (5)$$

$$2FeS + 7Cl_2 \longrightarrow 2FeCl_3 + 4SCl_2 \qquad (6)$$

$$2FeS_2 + 10SCl_2 \longrightarrow 2FeCl_3 + 7S_2Cl_2 \qquad (7)$$

$$S_2Cl_2 + 8H_2O + 5Cl_2 \longrightarrow 2H_2SO_4 + 12HCl \text{ (fast)} \qquad (8)$$

$$RH + S_2Cl_2 \longrightarrow RS_2Cl + HCl \text{ (slow)} \qquad (9)$$

$$FeS_2 + 7Cl_2 + 8H_2O \longrightarrow FeCl_2 + 2H_2SO_4 + 12HCl. \qquad (10)$$

These reactions are exothermic in nature and occur favorably at moderate temperature. The overall chlorine requirement to convert organic sulfur to sulfonate sulfur and sulfate sulfur is approximately 3 moles of Cl_2 and 4 moles of Cl_2, respectively, per mole of organic sulfur and 3.5 moles of Cl_2 per mole of inorganic sulfur.

In the presence of water and at a temperature (i.e., >50°C) higher than room temperature, the S_2Cl_2 formed from FeS_2 chlorination is readily converted to HCl and H_2SO_4. However at room temperature and without adequate moisture, this reaction is slow, and S_2Cl_2 may react with organic compounds to form organo-sulfur compounds. On the other hand, in an organic solvent at a slightly elevated temperature, the rate of chlorination of coal is slower than in aqueous media at room temperature. Also organic solvent media give a greater degree of structural loosening of coal and consequently more organic sulfur may be removed with a lesser degree of chlorination. Structural loosening of coal by the action of the organic solvent will make chlorine more accessible to sulfur compounds. High chlorine solubility in an organic solvent may also be advantageous for desulfurization. Moreover, an organic solvent may dissolve some of the organo-sulfur compounds. Chlorination of the coal matrix is mainly a substitution reaction, and hydrogen chloride is evolved as a product. Chlorine in coal chlorinated under mild conditions can be removed completely as hydrogen chloride by heating at 300°-550°C. Chlorination at high temperature and pressure results in coal which is difficult to dechlorinate.

After chlorinolysis coal becomes non-volatile, as indicated by no tar formation on heating up to 550°C. Only HCl gas is evolved on heating of chlorinated coal. Also chlorinated coal becomes non-caking and non-swelling upon heating. It appears reasonable to infer that the carbon present in structures normally giving rise to the formation of tar, which may be assumed to be hydroaromatic or other readily dehydrogenated structures, is retained in the solid residue by aromatization, crosslinking, or other forms of condensation reactions.

Hydrolysis. Chlorinated coal is hydrolyzed to produce hydrochloric acid according to the following reaction:

$$RCl + H_2O \longrightarrow ROH + HCl \qquad (11)$$

where R represents a hydrocarbon group in coal.

The sulfur converted to sulfates or sulfonate is water soluble and is leachable by water washing at 60°C for retention times up to 2 hr in a stirred reactor.

Dechlorination. After hydrolysis, the coal is dechlorinated by heating in steam or an inert gas atmosphere. This can be accomplished easily because of chlorination at low temperature. The possible reactions during dechlorination are:

1) in an inert gas atmosphere:

$$RH + R'Cl \longrightarrow RR' + HCl \qquad (12)$$

and

2) in steam atmosphere:

$$RCl(s) + H_2O(g) \longrightarrow ROH(s) + HCl(g) \qquad (13)$$

This reaction is endothermic and proceeds favorably at a moderately high temperature (300°-500°C).

According to the literature (16) steam will assist pyritic sulfur removal. Dechlorination in a steam atmosphere proceeds by substitution of -Cl in chlorinated coal by -OH and possibly -H groups from H_2O. No loss of heating value is experienced by the processed coal when dechlorinated in a steam atmosphere.

Laboratory Processing

The laboratory processing of coal for desulfurization by chlorinolysis is depicted in Figure 1. Chlorine gas is bubbled through a suspension of powdered (100-200 mesh) moist high sulfur coal in methyl chloroform at 74°C and atmospheric pressure for 1 - 4 hr. The coal slurry is distilled for solvent recovery. The chlorinated coal is hydrolyzed with water at 50°-70°C for 2 hr and then filtered. The coal filter cake is dried and dechlorinated by heating at 300°-500°C in a steam or vacuum atmosphere for approximately 1 hr. Analysis of the original and processed coal was made for organic, pyritic, and sulfate sulfur as well as an ultimate analysis of the coal including an analysis for trace metal removal by the desulfurization process.

Coal. Two high-sulfur coals were used for experimental studies. The Illinois No. 5 high-volatile bituminous coal from the Hillsboro mine had 4.77% total sulfur content. The other high-volatile bituminous coal was a Kentucky No. 9 coal from Hamilton, Kentucky. Proximate analyses of these two coals are given in Table I.

TABLE I. Proximate Analysis Data of Two Tested Coals

Coal	Fixed Carbon	Volatile Matter	Moisture	Ash
Illinois No. 5	42.74%	36.35%	9.88%	11.03%
Kentucky No. 9	52.45%	35.0%	4.49%	8.06%

Apparatus. The chlorination unit consists of a 1000-ml glass kettle heated by a constant temperature bath and equipped with a stirrer, a chlorine bubbler made up of fritted glass discs, a slurry sampling line, a reflux condenser suitable for complete total reflux of the solvent, and a gas holder. The hydrolyzer consists of a 2000-ml glass kettle equipped with a heating mantle, a temperature controller, a stirrer, a total reflux condenser, a slurry sampling line, and a gas holder.

The dechlorination unit consists of a rotary dryer (1/2 in. i.d. x 24 in. long) equipped with a heater and a temperature controller. The dryer is made up of a steel shell with a protective lining that includes acid resistant brick suitable for operations up to 600°C and at elevated pressure. The dryer has provision for vacuum, steam, or inert gas operation. The coal sample is heated inside the rotary dryer in vacuum, steam, or inert gas atmosphere. The outgoing gases pass through cold traps to a gas holder. For example, steam from a steam generator flows over the hot coal at 300°-450°C inside the dryer for less than 1 hr.

Results and Discussion

Experimental data obtained with a high-sulfur (4.77%) bituminous coal from Hillsboro, Illinois are presented in Table II. Chlorination was carried out in a stirred reactor operating at atmospheric pressure and 74°C. Chlorine was bubbled into the reactor for 1 hr. Powdered coal between 100 and 150 mesh was used with 50% water and a methyl chloroform-to-coal ratio of 2/1. The original sulfur content of the coal consisted of 1.89% pyritic, 2.38% organic, and 0.5% sulfate sulfur. After hydrolysis and water washing of the chlorinated coal at 60°C for 2 hr in a stirred reactor with excess water, the pyritic sulfur was 0.43%, organic sulfur was 0.72%, and sulfate sulfur was 0.35%. It is assumed that additional water washing would remove the sulfate sulfur completely. The hydrolysis solution contained mainly HCl, H_2SO_4, and other soluble sulfate and chloride compounds. The expected overall sulfur removal was 76% with a reduction of coal sulfur from 4.77% to 1.15%. Results of all the experiments with this coal indicate that removal of up to 70% organic sulfur, up to 90% pyritic sulfur, and 76% total sulfur have been achieved by current chlorinolysis procedures. Although the EPA standards for residual sulfur (~0.7%) have not been met for this high-sulfur coal, the guidelines should be met by further improvements in the process.

Chlorinated coal had an average of 22% chlorine before and 11% after the hydrolysis step. Dechlorination of the washed coal was carried out (Table III) at 500° and 550°C under vacuum for 1 hr. The residual elemental chlorine content was 0.06 wt % at 550°C and 0.15-0.3 wt % at 500°C. Furthermore, heating the powdered coal to 450°C for 1 hr under a steam atmosphere gave a residual elemental chlorine content of 0.064 wt %. These are compared with the original elemental chlorine content of 0.14%. Therefore, dechlorination of the coal is not expected to be a problem.

TABLE II.   Preliminary Chlorinolysis Data for Bituminous Coal
(Hillsboro, Ill.) Desulfurization

Sulfur Form	Raw Coal (% Sulfur)[a]	Treated Coal (% Sulfur)[a]	Sulfur Removal (%)	
			Achieved	Expected
Pyritic	1.89	0.43	77	77[b]
Organic	2.38	0.72	70	70
Sulfate	0.50	0.35	30	100[c]
Total	4.77	1.50	69	76

[a]Analyses by Galbraith Laboratories, Inc., Knoxville, Tenn.

[b]Up to 90% pyritic sulfur removal has been achieved in other conditions.

[c]Additional water washing should remove 100% of sulfate.

•For a 3% sulfur bituminous coal with same sulfur distribution as this coal, the 76% removal level may meet the EPA standard of 0.7% sulfur.

TABLE III.   Preliminary Dechlorination Data (Hillsboro, Ill.
Bituminous Coal)

Coal	Dechlorination			Elemental Chlorine (wt %)
	Temp (°C)	Time (hr)	Atmosphere	
Raw coal	–	–	–	0.14
Chlorinated coal (after hydrolysis)	–	–	–	11.0
Dechlorinated coal	450	1	steam	0.064
Dechlorinated coal	500	1	vacuum	0.15–0.30
Dechlorinated coal	550	1	vacuum	0.06

The kinetic data for chlorination and desulfurization of Illinois bituminous coal (-100 mesh) are presented in Figure 2. The initial rate of chlorination is very fast. The chlorine content of coal is 23% in 1/2 hr and then slowly increases to 26% within the next 1½ hr. Within the 1/2-hr period most of the pyritic sulfur and a portion of the organic sulfur are converted to sulfate sulfur. In the next 1½-hr period pyritic and organic sulfur are slowly converted to sulfate sulfur. Based on the sulfur balance, the gain in sulfate sulfur is equal to the combined reduction of pyritic and organic sulfur. The above reactions extend to the hydrolysis period. The overall sulfate compounds, converted from organic sulfur, pyritic sulfur and original sulfate sulfur, produced either directly or indirectly through sulfonate are removed from coal in the hydrolysis step as indicated by an analysis of the hydrolysis solution.

Heating values of the original coal and of the final treated coal were 11052 Btu/lb and 11,340 Btu/lb, respectively, on an as-received basis. The final value represents a slight increase over the original value. The increase is attributed to a difference in moisture content, etc.

Trace element concentrations found in the original Hillsboro, Ill., bituminous coal are presented in Table IV along with the trace element concentrations after chlorination and hydrolysis. Preliminary data indicated appreciable reductions of Pb, Va, P, Li, Be, and As. A number of these trace elements can pose serious environmental pollution problems unless their concentrations are reduced by the coal pretreatment.

Results of a chlorinolysis experiment with the addition of 0.5% $AlCl_3$ indicated that $AlCl_3$ did not have any additional effect on the removal of organic sulfur from coal. Coal contains sufficient iron to form about 1 - 2% $FeCl_3$ during chlorinolysis. This quantity of iron in the form of $FeCl_3$ may be adequate to catalyze ionic chlorination reactions.

Results of chlorinolysis experiment with dried coal (no moisture) and 0.5% $AlCl_3$ showed only 18.5% organic sulfur removal. The presence of moisture has a significant effect on aiding the removal of organic sulfur from coal and establishes the requirement for a minimum moisture content in the coal. Water participates in the reaction of RSCl and S_2Cl_2 to form H_2SO_4, HCl, and ROH.

Experimental data obtained with a 2.90% sulfur bituminous coal from Hamilton, Kentucky, are presented in Table V. The sulfur in the coal was predominantly organic sulfur at 2.67% with only 0.08% pyritic sulfur and 0.15% sulfate sulfur. Powdered coal (-200 mesh) with 30% moisture was chlorinated at 74°C and atmospheric pressure for up to 4 hr using a methyl chloroform-to-coal ratio of 2/1. After hydrolysis and water washing at 60°C in a stirred reactor for up to 2 hr, the total sulfur was reduced from 2.90 wt % to 1.48 wt % for a 57% organic sulfur reduction.

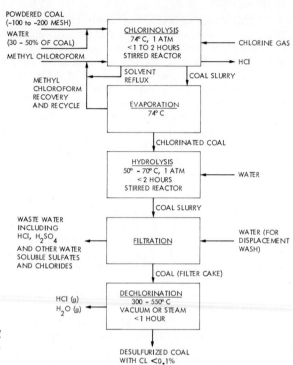

Figure 1. Laboratory coal processing diagram for desulfurization

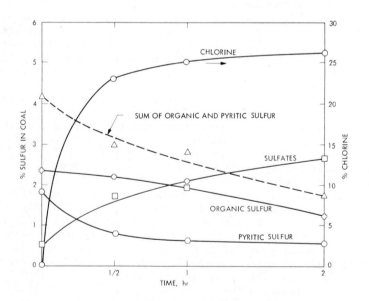

Figure 2. Sulfur and chlorine in coal during chlorination (74°C; 1 atm; coal, 35%; H₂O, 15%; CH₃CCl₃, 50%)

TABLE IV.   Results of Trace Metal Removal from Coal[a]

Elements	Original Coal[b] (ppm)	Treated Coal[b] (ppm)
As	2	1
Hg	0.6	< 0.5
Ti	476	460
Pb	18	4
Va	15	< 1
P	736	126
Se	< 1	< 1
Li	9	3
Be	3	1
Ba	38	30

[a] A high-sulfur bituminous coal with 11% ash, from Hillsboro, Ill. chlorinated at 74°C and atmospheric pressure for 1 hr, followed by aqueous leaching at 60°C and atmospheric pressure for 2 hr.

[b] Chemical analyses were conducted by the Galbraith Lab, Inc., Knoxville, Tennessee.

TABLE V.   Preliminary Chlorinolysis Data for Bituminous Coal (Hamilton, Ky.) Desulfurization

Sulfur Form	Raw Coal (% Sulfur)[a]	Treated Coal (% Sulfur)[a]	Sulfur Removal (%)	
			Achieved	Expected
Pyritic	0.08	0.03	62.5	62.5
Organic	2.67	1.16	56.5	56.5
Sulfate	0.15	0.29	0.0	100[b]
Total	2.90	1.48	51.0	59.0

[a] Analyses by Galbraith Laboratories, Knoxville, Tenn.

[b] 100% sulfate removal by added water wash.

Conclusion

A simple method of coal desulfurization by chlorinolysis of powdered moist coal in methyl chloroform at 74°C and 1 atm can remove up to 70% organic sulfur, up to 90% pyritic sulfur, and 76% total sulfur from a high-sulfur bituminous coal. After hydrolysis, the chlorinated coal is dechlorinated to 0.064% chlorine by heating under atmospheric pressure to 450°C in a steam atmosphere for less than 1 hr. The desulfurization method was also effective for another bituminous coal with high organic sulfur. The processed coal has reduced concentrations of toxic trace elements such as Pb, Va, P, Li, Be, As, etc.

This chlorinolysis method can be used as a cleaning process prior to direct combustion or gasification of coal for power generation. The hydrogen chloride recovered from the chlorination, dechlorination, hydrolysis, and distillation stages can be converted by the commercial Kel-chlor process (17,18) to chlorine for use in the process. The Kel-chlor process provides an economical conversion of HCl to chlorine. Choice of materials of construction will affect the process cost, which will be addressed in a later process paper. However, based on extensive discussions with people in the chlorination industry, it is expected that the technical and economical solutions for this process are available within the state-of-the-art technologies. A preliminary cost estimate of the desulfurization process (12,500 tons of coal/day) including the chlorine recovery unit (Kel-chlor process) indicates an overall process cost of $9 - $10 per ton of coal. This process offers high potential as a near-term solution for using high-sulfur coals as a source of clean energy.

Literature Cited

1. Friedman, S., Warzinski, R.I., Trans ASME, (1976) Paper No. 76-WA/APC-2.
2. Reggel, L., Raymond, R., Wender, I., Blaustein, B.D., Am. Chem. Soc., Div. of Fuel Chem., Preprint, (1970) Vol. 17, (1), 44-48.
3. Meyers, R.A., Hydrocarbon Process., (June 1975) 93-95.
4. Worthy, W., Chem. Eng. News (July 7, 1975) 24-25.
5. Macrae, J.C., Oxtoby, R., Fuel, Lond. (1965) 44, 395.
6. Pinchin, F.J., Fuel, Lond. (1958) 37, 293.
7. Ibid., (1959) 38, 147.
8. Oxtoby, R., Fuel, Lond. (1966) 45, 457.
9. Macrae, J.C., Oxtoby, R., Fuel, Lond. (1965) 44, 409.
10. Sun, S., Wu, M., AIME Annual Meeting, Preprint, (Dallas, Texas, Feb. 1974).
11. Imperial, G.R., Walker, P.L. Jr., Special Research Report SR-31, Pennsylvania State University, March, 1962.
12. Mukai, Shigeru et. al., Nenryo Kyokai-shi (1960) 48, 905.

13. Ganguli, P.S., Hsu, G.C., Gavalas, G.R., Kalfayan, S.H.,
     Am. Chem. Soc., Div. Fuel Chem., Preprint, (1976) vol. 21,
     (7), 118-123.
14. Kharasch, N., (ed.) "Organic Sulfur Compounds," Vol. I,
     Pergamon, New York, 1961.
15. Ezdakore, V.I., Tr. Uzb. Gos. Univ., Sbornik Rabot Khim,
     (1939) 15, 131.
16. Snow, P.D., Ind. Eng. Chem. (1932) 24, 903.
17. Schreiner, W.C., Cover, A.E., Hunter, W.D., van Kijk, C.P.,
     Jongenburger, H.S., Hydrocarbon Process. (Nov. 1974) 53, 151.
18. Process Eng. (April 1975) 7.

This paper presents the results of one phase of research carried
out at the Jet Propulsion Laboratory, California Institute of
Technology, under Contract No. NAS 7-100, sponsored by the
National Aeronautics and Space Administration.

Removal of Sulfur by Pyrolysis, Hydrodesulfurization, and Other Gas–Solid Reactions

Desulfurization and Sulfidation of Coal and Coal Char

G. J. W. KOR

U. S. Steel Corp. Research Laboratory, Monroeville, PA 15146

Desulfurization of Coal and Coal Char at Various Temperatures and Pressures

This work was undertaken to obtain a better understanding of the desulfurization of Illinois No. 6 coal and of char derived therefrom. In particular, the effects of temperature, pressure, and methane content of the gas on the rate of sulfur removal and the final sulfur content of the product were studied.

The desulfurization of high-sulfur coals, cokes, and chars has been the subject of many investigations in the past, those pertaining to the desulfurization of coke going back as far as the 1850s. In recent times the subject has gained importance because of the necessity of using large reserves of high-sulfur coal and of reducing the emission of sulfur-bearing gases in plants using coal or coke.

Experimental. Illinois No. 6 coal or char derived therefrom was used in all experiments; the size of the particles was between 12 and 18 mesh (average particle diameter ~1.3 mm). The coal was dried for 24 hr at 110°C before it was used. Two types of char were prepared by treating the dried coals in H_2 for 3 hr at 600° or 800°C. The total sulfur content and the amounts of the various forms of sulfur present in the coal and the char (prepared at 600°C) are given in Table I.

The desulfurization experiments, using either dried coal or one of the chars, were done in H_2, He, CH_4, and mixtures of H_2 and CH_4 for up to 3 hr at 600° and 800°C at pressures up to 10 atm. For each experiment 100–250 mg of sample contained in a platinum or nickel basket was suspended in the hot zone of a resistance furnace. The gas flow rate was 0.5 L(STP)/min in all cases. The samples were lowered into and pulled out of the hot zone as quickly as possible under a flow of He. At the end of an experiment the entire sample was analyzed for total sulfur by the combustion method (1).

In selected cases surface area measurements on partially desulfurized samples were made using the BET method. Use was also made of electron-probe analysis and optical microscopy, especially where the form of the sulfur was of interest.

Results and Discussion. The results for the desulfurization of dried coal in H_2, CH_4, and He at 600° and 800°C are shown in Figures 1 and 2. In all cases sulfur was rapidly lost the first 30 min. This initial rapid desulfurization is caused partly by the reduction of pyrite (FeS_2) to pyrrhotite (FeS) and partly by the loss of less stable organic sulfur.

The coal originally contained 0.52% sulfur as pyrite (Table I), present as particles with an average diameter 1-50 μ. Typical pyrite particles, as observed in the dry coal, are shown in Figure 3a and 3b. Electron-probe analysis showed that these particles have a composition approaching that of FeS_2 (Figure 3c).

The partial pressure of sulfur in equilibrium with FeS_2 and FeS is 1 atm at 690°C (2). Therefore, some decomposition of pyrite into pyrrhotite is expected in an inert atmosphere at 600°C. This is shown in Figure 3d, 3e, and 3f for a coal which was treated for 10 min in He at 600°C. The composition of the large porous particle in Figure 3d was close to that of pyrrhotite, as shown by Figure 3e, whereas the two smaller particles in Figure 3d had a composition between pyrite and pyrrhotite (Figure 3f).

Figure 3g is a micrograph of char prepared at 600°C, the chemical analysis of which is shown in Table I. This char was the starting material for subsequent desulfurization experiments. The porous particles in the center of the micrograph are pyrrhotite formed by the complete reduction of pyrite, as evidenced by the x-ray spectrum in Figure 3h. This observation is in keeping with the chemical analysis in Table I which showed that no pyritic sulfur was present in this char.

The observed gasification in CH_4 is the same at 600° and at 800°C whereas in He or H_2 the gasification is about 20% higher at 800°C than at 600°C. It is generally accepted that coal is carbonized in two stages (3). In the temperature range 350° - 550°C, the so-called primary devolatilization (not involving CH_4) takes place. The secondary gasification, involving mainly the release of CH_4 and H_2, begins at about 700°C. In the presence of CH_4, secondary gasification is therefore inhibited at 800°C and involves mainly the primary devolatilization equal to that observed at 600°C

Although two types of char were used in the desulfurization experiments, only the experimental results pertaining to the

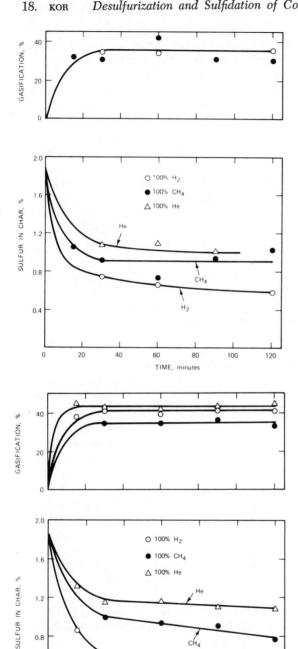

Figure 1. Desulfurization and gasification of dried Illinois No. 6 coal in H_2, CH_4, and He at 600°C and 1 atm

Figure 2. Desulfurization and gasification of dried Illinois No. 6 coal in H_2, CH_4, and He at 800°C and 1 atm

char prepared at 600°C are presented here. The results obtained
for the char prepared at 800°C are similar and will not be
presented in detail.

The effect of pressure and composition of the H_2-CH_4 mixture
and temperature on the desulfurization rate of char prepared at
600°C is shown in Figures 4, 5 and 6. Lower sulfur contents and
higher desulfurization rates are favored by an increase in p_{H_2}
and temperature. The presence of CH_4 inhibits desulfurization.

In gas mixtures containing more than 25% CH_4, gasification
ceased after about 30 min of reaction time to a plateau (Figure
5). The gasification corresponding to this plateau is shown as
a function of the percentage of CH_4 in the gas in Figure 7. For
the char prepared at 600°C, gasification during desulfurization
at 800°C (TS) decreased considerably with increasing CH_4 percentage in the gas. Gasification for the char prepared at 800°C only
slightly depended on the amount of CH_4 in the gas, irrespective
of the temperature of desulfurization.

As seen from these results, desulfurization of coal char
takes place in two distinct stages. The first stage shows a
simultaneous rapid desulfurization and gasification. During
the second stage sulfur is removed more slowly, practically
independent of the extent of any further gasification. The
observed initial rapid loss of sulfur together with the initial
rapid gasification suggests that a relationship may exist between
the initial fractional removal of sulfur, $(\Delta S/S_o)_i$, and that of
carbon, $(\Delta C/C_o)_i$.

The data in Figure 8 show the relative sulfur removal after
about 15 min reaction time as a function of the relative carbon
loss incurred during this time. A similar relationship was
observed for the char prepared at 800°C. It is seen from Figure
8 that for each desulfurization temperature (TS) the data points
corresponding to various experimental conditions (e.g., total
pressure, CH_4 content) form a curve, indicating a relationship
between $(\Delta S/S_o)_i$ and $(\Delta C/C_o)_i$.

Further examination of the data in Figure 8 shows that the
same functional relationship between $(\Delta S/S_o)_i$ and $(\Delta C/C_o)_i$
exists, irrespective of the temperature at which desulfurization
took place. This is shown in Figure 9 where the open circles
represent all the data points in Figure 8. Also in Figure 9
are the results obtained for the char prepared at 800°C, which
show a similar relationship between $(\Delta S/S_o)_i$ and $(\Delta C/C_o)_i$;
however, the slope is steeper than observed for the char prepared
at 600°C. The higher the char preparation temperature, the more
gasification and desulfurization takes place during charring.

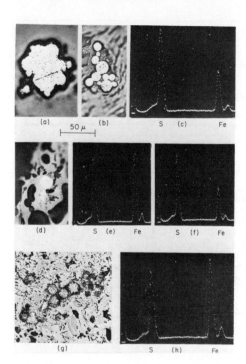

Figure 3. Micrographs and electron probe analysis for coal and char. (top) Original dry coal; (middle) coal, treated in He at 600°C for 10 min; (bottom) char (3 hr in H₂ at 600°C).

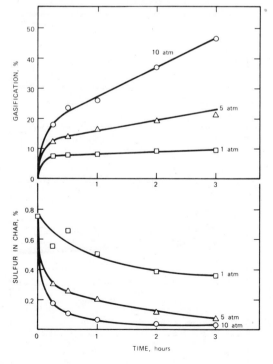

Figure 4. Desulfurization and gasification of Illinois char (prepared at 600°C) in H₂ at indicated pressures at 800°C

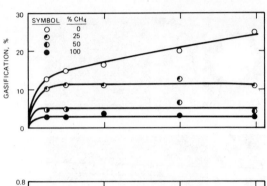

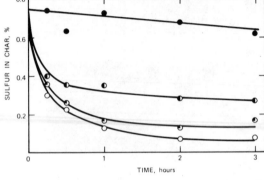

Figure 5. Desulfurization and gasification of Illinois char (prepared at 600°C) in H_2–CH_4 gas mixtures of 5 atm at 800°C

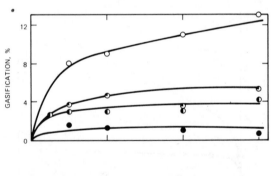

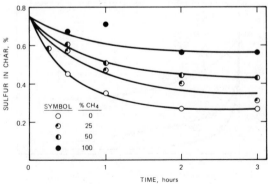

Figure 6. Desulfurization and gasification of Illinois char (prepared at 600°C) in H_2–CH_4 mixtures of 5 atm at 600°C

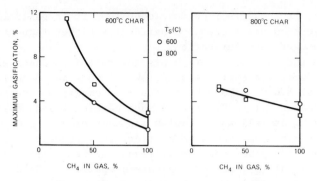

Figure 7. Final level of gasification as a function of the amount of CH_4 in H_2–CH_4 mixtures for two types of char, during desulfurization at 600° and 800°C (T_s)

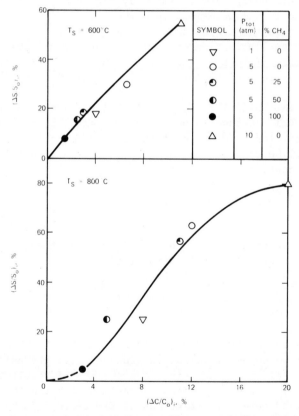

Figure 8. Relation between initial loss of sulfur and carbon (after 15 min reaction time) for a char prepared at 600°C and subsequently desulfurized under indicated conditions at $T_s = 600°$ or 800°C

Therefore, it should not be concluded that it is generally more advantageous to use a char, prepared at a higher temperature, for subsequent desulfurization.

Also included in Figure 9 are the data from Jones et al.(4) who desulfurized a char derived from Illinois No. 6 coal, prepared at 870°C. The desulfurization temperature varied between 704° and 1010°C, the pressure between 1 and ~8 atm, and they used H_2 as well as equimolar mixtures of H_2 and CH_4. Although their data show some scatter, it is concluded that there is a relationship between $(\Delta S/S_o)_i$ and $(\Delta C/C_o)_i$. The data of Batchelor et al.(5), who used a char prepared at 500°C from a Pittsburgh seam coal, are also shown in Figure 9. Desulfurization took place in H_2-H_2S mixtures (p_{H_2} between 1 and 11 atm) at temperatures varying between 650° and 880°C. Also for these data, a relationship between $(\Delta S/S_o)_i$ and $(\Delta C/C_o)_i$ is observed. Thus the functional relationship between $(\Delta S/S_o)_i$ and $(\Delta C/C_o)_i$ depends only on the temperature at which the char was prepared. Subsequent desulfurization of a given char can only be achieved at the expense of loss in carbon, the extent of which is determined by the appropriate functional relationship depicted in Figure 9.

After the initial rapid drop in sulfur content of the char, a more gradual decrease is observed(Figures 4, 5, and 6). Assuming that for this stage of the process, the desulfurization reaction may be described by a first-order reaction relative to the sulfur content of the char, then

$$\frac{dS}{dt} = - k_S S \tag{1}$$

where S is the concentration of sulfur at time t and k_S a rate constant. Integration of Equation 1 gives:

$$\log \frac{S}{S_o} = - k_S t \tag{2}$$

where S_o is the sulfur concentration after 15 min of reaction time, after which desulfurization proceeds more gradually.

Because of the scatter in the experimental results, it was not necessary to treat the results obtained for the two chars separately. Figure 10, depicting the first-order plots, therefore represents the averages for both types of char. Log (S/S_o) is a linear function of time within the scatter of the data.

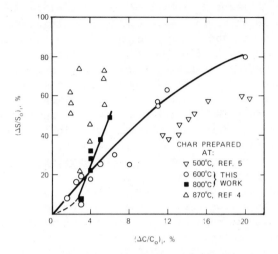

Figure 9.   *General relationship between initial loss of sulfur and carbon for two types of char, showing that the data for various experimental conditions for a given char are represented by the same function*

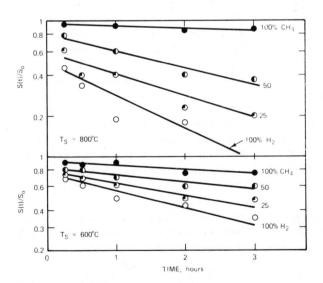

Figure 10.   *Desulfurization of char in mixtures of H_2 and CH_4 at 5 atm and indicated temperatures, as a first-order reaction*

The rate constant k_s, obtained from the slopes of the lines in Figure 10, is shown in Figure 11A as a function of the concentration of H_2 in the H_2-CH_4 mixture for the desulfurization experiments at 5 atm at 600° and 800°C. Although the equilibrium concentrations of CH_4 in H_2-CH_4 mixtures at 5 atm are 56% and 16% at 600° and 800°C, respectively (2), no measurable weight increase of the char was recorded after treatment in CH_4 at 800°C. This indicates that CH_4 did not dissociate to any measurable extent under the present experimental conditions. Therefore, it may be assumed that the partial pressure of H_2 prevailing during the desulfurization experiments in H_2-CH_4 mixtures was the same as that in the ingoing mixture.

On this basis, Figure 11B was plotted, supplemented with some data obtained from desulfurization experiments in 100% H_2 at 1 and 5 atm. The rate constant pertaining to the second stage of desulfurization in 100% H_2 is the same as that in H_2-CH_4 mixtures, although in 100% H_2 gasification continues in the second stage (Figures 4, 5 and 6). This suggests that desulfurization and gasification are interrelated in the initial stages only; in the second stage the desulfurization is independent of gasification.

To explain these observations, it is suggested that the initial loss of carbon—which is accompanied by a simultaneous loss of (mainly organically bound) sulfur—creates new pores providing better access for the reducing gas to the pyrrhotite particles embedded in the char. This is supported by the observed change in surface area of the char during desulfurization. The initial surface area of char prepared at 600°C is about 2 m^2/g. The change in surface area is most pronounced during the first hour of desulfurization, particularly at 800°C (Figure 12). The pore surface area increases with increasing temperature and pressure and, hence, so does the amount of gasification.

A char, prepared at 600°C and subsequently desulfurized for 2 hr in 5-atm H_2 at 800°C, was analyzed for the various forms of sulfur present in the product. The analysis showed that of the total sulfur content of 0.16%, about 0.11% was pyrrhotite and about 0.05% was organic sulfur. The micrograph in Figure 13a for this partially desulfurized product shows three types of particles: bright, greyish colored, and two-phase particles partly bright and partly grey. The composition of the greyish particles varies somewhat but nominally approaches that of pyrrhotite (Figure 13b and 13c). The bright particles are iron (Figure 13d) formed by the complete reduction of pyrrhotite.

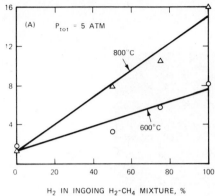

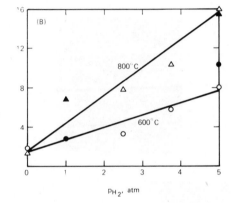

Figure 11. *First-order rate constants for desulfurization at indicated temperatures shown as a function of (A) H₂ content of gas mixture (5 atm) and (B) partial pressure of H₂ (solid symbols represent data for 100% H₂)*

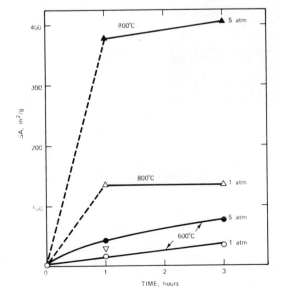

Figure 12. *Surface area (SA) of char after treatment in H₂ at indicated temperatures and pressures*

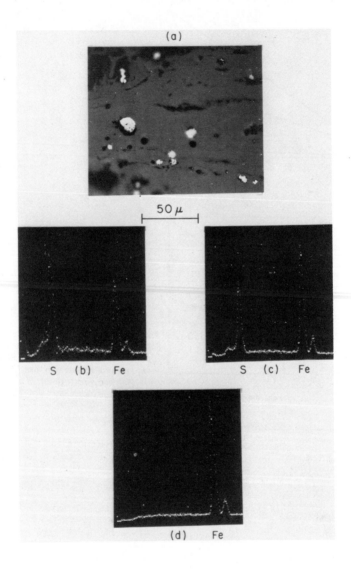

Figure 13. Optical micrograph and electron analysis of char,
treated in $p_{H_2} = 5$ *atm, 800°C for 2 hr, showing the presence of*
pyrrhotite and reduced iron

It is concluded from these observations that most of the sulfur is in the form of pyrrhotite during the later stages of desulfurization. The overall rate of desulfurization during this stage is thus expected to be mainly governed by the slow reduction (6) of pyrrhotite in H_2.

Conclusions. It was found that the desulfurization of coal char in mixtures of H_2 and CH_4 takes place in two distinct stages. In the first stage rapid desulfurization is accompanied by gasification. These two processes were shown to be interrelated, the relationship depending on the char preparation temperature only. The second stage of desulfurization proceeds at a much slower rate and is being controlled by the slow reduction of pyrrhotite to iron.

Sulfidation of Coal Char and Synthetic Chars

Desulfurization of coal and coal char in hydrogen results in evolution of H_2S. Depending on the process, the H_2S is either entirely or partly removed and recirculated. The work, described in Part II, was undertaken to obtain a better understanding of the interaction between chars and gas mixtures containing H_2S.

A literature survey indicated that no investigations have been made of the sulfidation of carbonaceous materials, including chars, in gas mixtures of H_2 and H_2S such that the sulfur potential was systematically varied.

Experimental. Char from two different sources, prepared under a variety of conditions, was used in this work. The preparation conditions are summarized in Table II.

The preparation of char from Illinois No. 6 coal was described in the first section of this chapter under "Experimental". Char from ash-free filter paper (0.008% ash) was prepared by charring the paper, contained in a high-purity alumina boat, in an atmosphere of dry He at 600° or 900°C for 3 hr. After the paper was charred, the boat was pulled to the cool end of the reaction tube where it slowly cooled. Subsequently, the char was quickly transferred to a desiccator where it was stocked. The chars that were further treated in He at 1250° and 1500°c for 24 and 96 hr, respectively, were taken from this stock.

The sulfidation experiments were done in a vertical furnace with the H_2–H_2S mixture entering at the bottom of the reaction tube. The H_2–H_2S-ratio in the gas was adjusted by using the usual arrangement of capillary flow meters. In most of the experiments a gas flow rate of 30 cm^3(STP)/min was used.

For each sulfidation experiment about 100 mg of char was placed in an alumina tray. In most cases the sulfidizing treatment was 1 hr after which the sample was quickly pulled up to the cool top of the reaction tube while the He was kept flowing. The cooled sample was then transferred to a desiccator. Sulfur in the char was analyzed by the combustion method (1).

In some selected cases about 750 mg of char was sulfidized, and afterwards analyzed for oxygen by the neutron-activation method. The surface areas of the chars used were determined by the BET method.

Results and Discussion. The partially desulfurized coal char (Table II) was sulfidized at 600°, 800°, and 900°C in H_2-H_2S containing 0-100% H_2S. The results are given in Figure 14, in which the sulfur content of the char after 1 hr reaction time is plotted vs. the percentage of H_2S in the gas at the experimental temperature, denoted by $(\%H_2S)_T$. Similar results were obtained after sulfidation for 4 hr.

In calculating $(\%H_2S)_T$ from the percentage of H_2S in the ingoing gas (at room temperature), due allowance should be made for the partial dissociation of H_2S at higher temperatures. The equilibrium H_2S percentage of the gas at the reaction temperature was calculated from the available thermodynamic data(2).

The present results may be compared with those obtained by Polansky et al.(7) who treated coke in H_2S-N_2 mixtures containing 4.2 and 8.8% H_2S. Their results show no pronounced difference in the extent of sulfur absorption at 800° and at 900°C.

Of special interest is the absorption of sulfur at 900°C in H_2-H_2S mixtures, containing less than 2% H_2S, by previously desulfurized coal char (from Illinois coal). This coal char was produced by desulfurization of a char that originally contained 0.13% sulfur as pyrrhotite; this is equivalent to about 0.23% iron. Upon sulfidation of this desulfurized coal char, pyrrhotite is expected to form when the sulfur potential is sufficiently high. This is illustrated in Figure 15, where a sudden rise in the amount of sulfur absorbed is observed when $(\%H_2S)_T > 0.3$. This is in good agreement with the value calculated from the thermodynamic data for the iron/pyrrhotite equilibrium(2) at 900°C. Sulfur content increases about 0.20% at the "break point" in the absorption curve, in good agreement with the estimated amount of iron present in the char.

The sulfur absorption curves, depicted in Figure 14, have the general character of absorption isotherms. However, proper interpretation of these results is hindered by the presence of

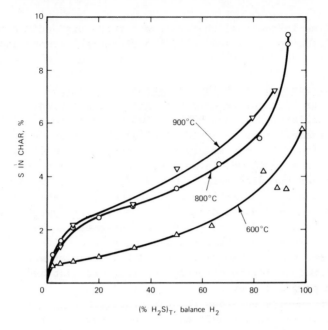

Figure 14. Sulfidation of previously desulfurized char from Illinois No. 6 coal at indicated temperatures

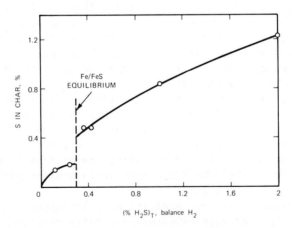

Figure 15. Sulfidation of previously desulfurized char at 900°C in H_2–H_2S mixtures for low percentages of H_2S

impurities in the coal char. It was therefore decided to study
the sulfidation of essentially impurity-free carbons and synthetic
chars.

Sulfidation of Synthetic Carbons and Chars. Granulated
samples of high-purity electrode graphite and pyrolitic graphite
were equilibrated with a 50% H_2, 50% H_2S mixture at 1000°C for
1.5 hr; there was no detectable sulfur absorption in either form
of graphite. In another experiment samples of electrode graphite
and pyrite were placed in separate parts of an evacuated silica
capsule and annealed for 20 hr at 650°. This corresponds to a
partial pressure of sulfur vapor of about 0.14 atm, as estimated
from the thermodynamic data for the pyrrhotite/pyrite system(2).
After this treatment, no sulfur was detected in the graphite.

The present observations generally agree with those of
Wibaut and van der Kam(8) who found that even at sulfur pressures
above atmospheric, no detectable sulfur was absorbed by either
diamond powder or Ceylon graphite.

The results obtained for the synthetic chars are given in
Figures 16 and 17 for 600° and 900°C, respectively. In most
experiments the reaction time was 1 hr; however, some samples
were sulfidized for up to 3 hr. These samples absorbed essentially
the same amount of sulfur as those sulfidized for 1 hr. Moreover,
equilibrium could be reached from both sides. For example,
filter-paper char (prepared at 900°C) which was first sulfidized
in a 50% H_2, 50% H_2S mixture at 600°C to a final sulfur content
of ~1% could subsequently be partially desulfurized in a 90%
H_2, 10% H_2S mixture to yield a final sulfur content of 0.4%.
This is essentially the same as the sulfur content after direct
sulfidation of the char in the same gas mixture. Similar
observations were made at 900°C, indicating that the absorbed
sulfur is in equilibrium with the gas and that the process of
sulfur uptake is reversible.

The surface areas of the chars (Table II) are indicated in
Figures 16 and 17. A char with a larger surface area has, in
general, a larger capacity for sulfur absorption. The results
obtained for coal char are also shown in Figures 16 and 17 for
easy comparison. Coal char and filter-paper char (prepared at
600° and 900°C) have about the same surface areas and absorb
similar amounts of sulfur.

The results of x-ray analysis of the various chars used in
this investigation were compared in a qualitative way with the
data reported by Turkdogan et al.(9) as shown in Table III
together with estimated mean crystallite sizes. It was mentioned

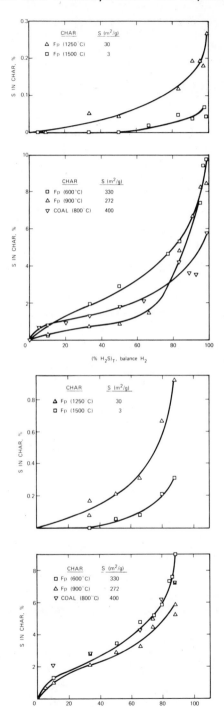

Figure 16.   Sulfidation at 600°C in H_2–H_2S mixtures of various chars, having indicated pore surface area (S)

Figure 17.   Sulfidation at 900°C in H_2–H_2S mixtures of various chars, having indicated pore surface area (S)

Table I. Forms of Sulfur (in Wt.%) in Dried Coal
          and Derived Char (3 hr, H_2, 600°C)

Form of Sulfur	Dried Coal	Coal Char
Pyrite	0.53	–
Sulfate	0.12	0.005
Sulfide	0.005	0.13
Organic	1.27	0.61
Total	1.93	0.75

Table II. Source, Conditions of Preparation, Initial Sulfur
           (S_i) and Oxygen (O_i) Contents, and Surface Area (SA)
           of the Various Chars Studied

Source	Preparation Conditions	Properties of the Char		
		$SA(m^2/g)$	$S_i(\%)$	$O_i(\%)$
Illinois Coal	5 atm H_2, 800°C for 3hr	400	0.05	–
Ash-free Filter Paper (0.008% ash)	1 atm He, 600°C, 3hr	330	0.00	4.0
	1 atm He, 900°C, 3hr	272	0.00	1.4
	1 atm He, 600°C, 3hr + 1 atm He, 1250°C, 24 hr	30	0.00	0.34
	1 atm He, 600°C, 3hr+ 1 atm He, 1500°C, 96hr	3	0.00	0.14

Table III. Qualitative Comparison of the Crystallinity of
            Filter–Paper Chars Used in This Work With the
            Crystallinity and Mean Crystallite Size of Some
            Carbons Investigated by Turkdogan et al.(9)

Type of Char and Preparation Temp. (°C)	Crystallinity from Turkdogan's Work	
	Qualitative Comparison From X-ray Analysis	Mean Crystallite Size (Å)
Filter paper (600)	between coconut charcoal and "vitreous" carbon	~10 – ~16
Filter paper (900)		
Filter paper (1250)	approaching vitreous carbon	~16
Filter paper (1500)	same as vitreous carbon	~16

before that graphitized electrode graphite, which has a mean crystallite size of 500 Å, did not absorb sulfur whereas the nongraphitized chars did. Thus, it is concluded from these results that the ability of a given char or carbon to absorb sulfur is determined first by its state of crystallinity. In poorly graphitized or nongraphitized carbons, the amount of absorbed sulfur increases with increasing pore surface area.

In view of the strong affinity between sulfur and oxygen, an attempt was made to investigate the effect of oxygen on the sulfur absorption by char. The initial oxygen content of synthetic chars is shown as a function of the surface area in Figure 18. It is seen that the initial oxygen concentration is a strong function of the surface area and hence the temperature at which the char was prepared (Table II).

To study the change in oxygen concentration after sulfidation of the chars, a series of special experiments was conducted; the results are summarized in Table IV. The oxygen content after sulfidation was independent, within the analytical error, of the ratio $(p_{H_2S}/p_{H_2})_T$ in the gas. An increase in the sulfidation temperature resulted in a lower oxygen content in the char, particularly if the char had a larger surface area (Figure 18).

In all cases investigated, the chars with large surface areas contained more oxygen. Because of the interdependence of surface area and oxygen content, it is difficult to separate their effects on the capacity of a given char for sulfur absorption. However, some indication of the influence of oxygen on the sulfur absorption may be obtained from the results shown in Table IV. For instance, the surface areas of filter–paper char prepared at 600° and 900°C were 330 and 272 m^2/g, respectively, a difference of about 20%. After sulfidation at 600°C in a gas of high sulfur potential, the oxygen contents of these chars differed by about a factor of two, yet the difference in sulfur absorbed was not more than about 10%. These findings indicate that the influence of oxygen on the sulfur absorption is probably secondary.

In this context the work of Hofmann and Ohlerich(10) should be mentioned. They treated sugar charcoal in dry O_2 at about 500°C to obtain a char containing about 10% oxygen as surface complexes. Upon sulfidation of this oxygenated char in S_2 at 600°C, they found that the amount of sulfur taken up was equal to that absorbed by a char which was not previously activated in oxygen. Hofmann and Ohlerich concluded, as did Hofmann and Nobbe(11), that the amount of sulfur absorbed by char depends only on its surface area.

Some of the filter-paper chars used in this work were analyzed for hydrogen and nitrogen. The results are summarized in Table V, which shows that the major impurities in filter-paper char are oxygen and hydrogen. The oxygen and hydrogen contents decrease with increasing temperature of char preparation while the nitrogen content remains essentially constant.

The shape of the curves in Figure 19, in which the amount of sulfur in some synthetic chars is shown as a function of $(p_{H_2S}/p_{H_2})_T$, strongly suggests absorptive behavior. Hayward and Trapnell(12) give examples of typical absorption isotherms and note that chemisorption normally gives rise to isotherms of this general form.

Although a treatment of the present results in terms of idealized absorption isotherms is open to criticism, an attempt will nevertheless be made to treat the results accordingly. It will be shown that such a treatment leads to results which may be considered reasonable.

Assuming that the chemisorbed sulfur forms an ideal monolayer and that each chemisorbed species occupies a single site, application of the ideal Langmuir isotherm gives(12):

$$a = \frac{\theta}{B(1 - \theta)} \tag{1}$$

where a is the activity of the chemisorbed species, θ the fractional coverage, and B a temperature-dependent parameter containing the heat of chemisorption of sulfur. The fractional coverage $\theta = v/v_m$, where v is the volume of chemisorbed sulfur (STP) per gram of char and v_m the volume giving a complete monolayer of sulfur on the surface of the char. The surface area, S m^2/g, is related to v_m by the following expression:

$$S = \frac{v_m}{22414} NA \cdot 10^{-20} \tag{2}$$

where N is Avogadro's number and A the cross-sectional area of an absorbed species in A^2. The volume, v, of the chemisorbed sulfur is obtained from the measured sulfur concentration as follows:

$$v = \frac{22414}{3200} (\%S) \tag{3}$$

Table IV. Oxygen and Sulfur Contents of Filter-Paper Chars after Sulfidation at 600° and 900°C for 1 hr in Gases of Various Sulfur Potential

Sulfidation Temperature (°C)	Preparation Temperature (°C)	$O_i(\%)^a$	$\left(\dfrac{P_{H_2S}}{P_{H_2}}\right)_T$	Sulfidized Char %O	%S
600	600	4.0	0.05	1.7	0.80
			61.5	1.6	9.50
	900	1.4	0.05	0.8	0.20
			61.5	0.8	8.40
	1250	0.34	0.05	0.2	0.03
	1500	0.14	0.05	0.1	0.02
			61.5	0.1	0.05
900	600	4.0	0.05	0.7	0.90
			7.6	0.6	7.50
	900	1.4	0.05	0.9	0.40
			7.6	0.6	5.80
	1500	0.14	0.05	0.1	0.01
			7.6	0.1	0.25

$^a O_i$ = initial oxygen content of the char.

Table V. Chemical Analysis of Filter-Paper Char in Relation to Preparation Temperature

Char Preparation Temperature (°C)	Composition (wt.%)			
	Carbon	Oxygen	Hydrogen	Nitrogen
600	93.9	4.0	1.9	0.02
900	97.0	1.4	0.6	0.06
1250	99.6	0.34	0.2	0.04

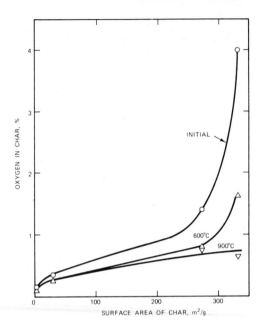

Figure 18. Oxygen concentration in synthetic chars as a function of surface area: initial, and after sulfidation at 600° and 900°C

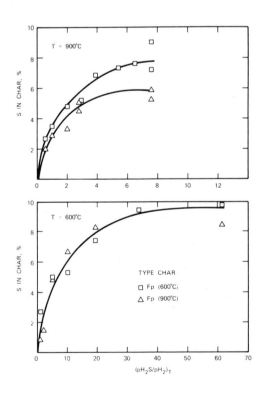

Figure 19. Sulfidation of various chars at 600° and 900°C, showing curves typical for chemisorption

Substituting for the sulfur activity $a = (p_{H_2S}/p_{H_2})_T$,
$\theta = v/v_m$, and making use of Equations 2 and 3, the following
expression is obtained from Equation 1:

$$\frac{1}{(\%S)} (p_{H_2S}/p_{H_2})_T = 1.87 \frac{A}{S} \left[\frac{1}{B} + (p_{H_2S}/p_{H_2})_T \right] \tag{4}$$

The experimental results plotted in accordance with Equation 4
are given in Figures 20 and 21 for 600° and 900°C, respectively.
The slope of each line should be proportional to $1/S$; in fact,
this is shown to be the case in Figure 22, in which the log
(slope) is depicted as a linear function of log S with a
theoretical slope of −1. From the intercept of this line with
the ordinate the value of A, the cross−sectional area of a
chemisorbed species, is calculated to be 17 $\overset{\circ}{A}^2$. This value is
to be compared with cross−sectional areas ranging from 10 to 50
A^2 as estimated from physical absorption data for a variety of
gases(12).

According to Equation 4, the intercepts of the lines in
Figures 20 and 21 with the ordinate should be proportional to
$1/S$. This is shown to be the case in Figure 23, in which log
(intercept) is depicted as a function of log S with a theoretical
slope of −1. From the intercepts of both lines with the ordinate,
together with the previously determined value of A, the temper−
ature−dependent parameter B is obtained. This parameter is
proportional to $e^{q/RT}$, where q is the heat of chemisorption of
sulfur on char. From the temperature dependence of B, the value
of q is estimated to be about −10 kcal/mole, a reasonable value
when compared with the heats of chemisorption of other gases on
carbon, as listed by Hayward and Trapnell(12).

The foregoing analysis, although of necessity oversimplified,
shows that the absorption of sulfur by synthetic chars is most
likely mainly governed by chemisorption. It is thus expected
that the surface area is an important parameter in determining
whether a given char or carbon is able to retain significant
quantities of sulfur. This is in agreement with earlier work
by Hofmann and Nobbe(11) and by Polansky et al.(7).

However, the surface area is not the only parameter to be
considered. For instance, electrode graphite in the unoxidized
state has a surface area of 1 m^2/g(9), approximately the same
as filter−paper char prepared at 1500°C (Table II). Yet, no
take−up of sulfur by electrode graphite was observed after
sulfidation in 50% H_2−50% H_2S at 1000°C. The mean crystallite

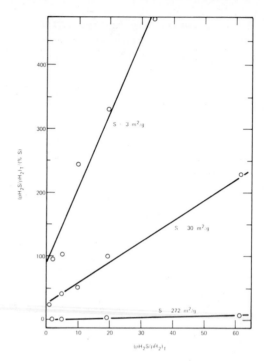

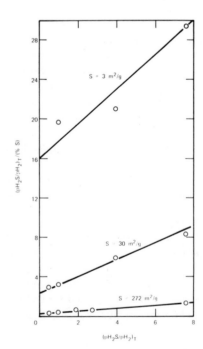

Figure 20. Langmuir isotherms show-
ing chemisorption at 600°C on chars
having indicated surface areas

Figure 21. Langmuir isotherms show-
ing sulfur chemisorption at 900°C on
chars having indicated surface areas

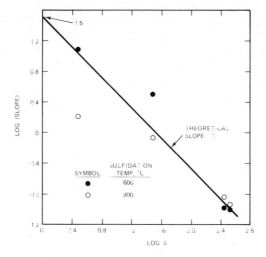

Figure 22. Slope of chemisorption isotherms at 600° and 900°C as a function of the surface area of the char

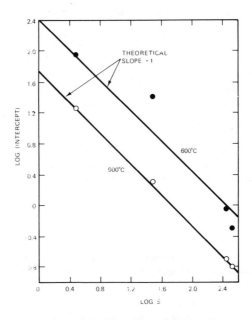

Figure 23. Intercept of chemisorption isotherms with ordinate as a function of surface area of the char, sulfidized at indicated temperatures

size, together with the x-ray analysis, of electrode graphite(9)
shows that its nature is graphitic and therefore more ordered.
Blayden and Patrick(13) concluded from their work that so-called
disordered carbons with small carbon layers (equivalent to mean
crystallite size) and many defects are better able to absorb
sulfur than the more crystalline and graphitic carbons. The
present findings are consistent with this viewpoint.

   Conclusions. In the absence of impurities such as iron,
the ability of chars or carbons to absorb significant amounts of
sulfur in a sulfidizing gas such as a H_2-H_2S mixture depends on
the state of crystallinity of the carbonaceous material. The
sulfur absorption decreases with increasing crystallite size.
In general, carbons having a mean crystallite size of about 15 Å
or less absorb significant amounts of sulfur when treated in
H_2-H_2S mixtures. For carbons with a given crystallite size, the
higher the pore surface area the higher is the amount of sulfur
absorbed.
   Sulfur absorption in high-purity chars, obtained from ash-
free filter paper, increased with increasing sulfur activity and
pore surface area of the char. This is in accord with the
Langmuir relation for chemisorption on single sites in an ideal
monolayer.
   It is concluded from the present experimental results that
sulfur is not accommodated in the three-dimensional lattice of
the carbon but is chemisorbed on the surface. However, such
chemisorption takes place only on the pore walls of nongraphitic
(poorly crystalline) carbons, of which chars are good examples.

Acknowledgment

   The author wishes to thank J. R. Cauley of this laboratory
for his assistance with the experiments.

Litreature Cited

1. "Sampling and Analysis of Coal, Coke, and By-Products,"
   pp. 84-86, Carnegie Steel Corp., Pittsburgh, 1929.
2. Kubaschewski, O., Evans, E. L., Alcock, C. B., "Metallurgical
   Thermochemistry," 4th ed., Pergamon, London, 1967.
3. van Krevelen, D. W., "Coal," Elsevier, Amsterdam, 1961.
4. Jones, J. F., Schmidt, M.R., Sacks, M.E., Chen, Y.C., Gray,
   C.A., Office of Coal Research, R&D Rept. 11, Suppl. to
   Final Rept., Jan. 9, 1967.

5. Batchelor, J. D., Gorin, E., Zielke, C. W., J. Ind. Eng. Chem., (1960) 52, 161–8.
6. Kor, G. J. W., Met. Trans. (1974) 5, 339–43.
7. Polansky, T. S., Knapp, E. C., Kinney, C. R., J. Inst. Fuel, (1961) 34, 245–6.
8. Wibaut, J. P., van der Kam, E. J., Rec. Trav. Chim. Pays-Bas, (1930) 49, 121–37.
9. Turkdogan, E. T., Olsson, R. G., Vinters, J. V., Carbon, (1970) 8, 545–64.
10. Hofmann, U., Ohlerich, G., Angew. Chem., (1950) 62, 16–21.
11. Hoffman, U., Nobbe, P., Ber. Dtsch. Chem. Ges., (1932) 65, 1821–30.
12. Hayward, D. O., Trapnell, B. M. W., "Chemisorption," 2nd ed., Butterworths, Washington, D. C., 1964.
13. Blayden, H. E., Patrick, J. W., Carbon, (1967) 5, 533–44.

19

Fluid-Bed Carbonization/Desulfurization of Illinois Coal by the Clean Coke Process: PDU Studies

N. S. BOODMAN, T. F. JOHNSON, and K. C. KRUPINSKI

U.S. Steel Research Laboratory, Monroeville, PA 15146

The Clean Coke Process combines both fluid-bed carbonization and hydrogenation/liquefaction to convert high-sulfur coal to low-sulfur metallurgical coke, chemical feedstocks, and to a lesser extent, liquid and gaseous fuels. The overall processing scheme, which has previously been described in detail (1), is illustrated by the sketch in Figure 1.

Briefly, run-of-mine coal is beneficiated and classified by conventional means and split into two feed portions: a sized fraction suited for fluid-bed processing and a fines fraction suited for high-pressure hydrogenation. The sized feed is dried and subjected to a mild surface oxidation in a nonpressurized bed fluidized with air-enriched flue gas. The dry, preoxidized feed is then carbonized in two stages, at 820°F (440°C) and 1400°F (760°C), in fluid-bed reactors operated at pressures up to 150 psig, to produce low-sulfur char, tar, and gas rich in methane and hydrogen. The fines fraction of the beneficiated coal, combined with run-of-mine coal, is dried, pulverized, and slurried with a process-derived oil. The slurry is then pumped to a pressure reactor and liquefied at 850°-900°F (455°-480°C) and 3000-4000 psig to produce liquids and C_1-to-C_4 hydrocarbon gases. Liquids from both operations are distilled to produce a light chemical oil, a middle oil for recycle to the hydrogenation reaction, and a heavy oil. The heavy oil, a soft pitch, is combined with the carbonization char and processed to make a low-sulfur metallurgical formcoke, currently in the form of pellets. Similarly, product gases from all operations are combined and processed to produce hydrogen for the hydrogenation operation, fuel, ethylene and propylene, sulfur, and ammonia. A detailed description of yield of chemical products and process economics has been presented previously (2).

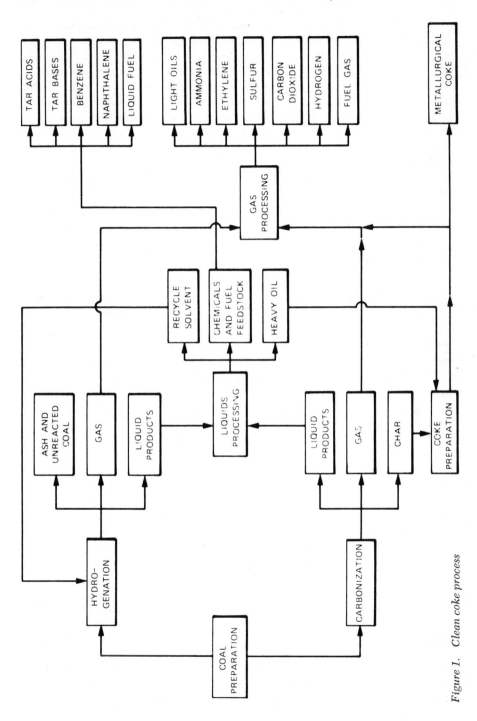

Figure 1. Clean coke process

This chapter presents the results obtained from sustained operation of the carbonization PDU (process development unit). These results confirm and extend the data obtained previously in bench studies (3). All tests were run with Illinois No. 6 seam coal containing 2-2.5% sulfur after preparation. From this were produced chars containing, generally, 0.6-0.7% sulfur; char containing as little as 0.2% sulfur was produced under the more severe reaction conditions. Reaction conditions investigated for their effect on char sulfur content included residence time, temperature, pressure, and H_2S concentration in the fluidizing gas. Data are also presented to show the weight distribution of materials into and out of the PDU system. The scale of the tests discussed in this paper is best illustrated by a description of the design and operation of the carbonization PDU.

Carbonization PDU

Figure 2 is a photograph of the carbonization PDU, in which the fluid-bed reactor occupies the second level; the feed vessels are on the top level, and the char receivers are at floor level. Two other vessels visible at floor level are liquid catchpots attached to the gas-to-gas heat exchanger (on the right) and the water-cooled exchanger (on the left). Construction details of the fluid-bed carbonizer are illustrated in Figure 3. The vessel, including top and bottom closures, is 9 ft 3 in. tall and is fabricated from 1-in.-thick Incoloy Alloy 800 to permit operation at 1500°F (815°C) and 150 psig. The lower 36-in. section of the reactor is the 10-in.-ID fluid-bed area; the expanded upper 36-in. section has a 20-in. ID to facilitate deentrainment of fine solids from the fluidizing gas. Feed enters the fluid bed by gravity flow through the feed pipe, positioned about 1 in. above the gas-distributor plate; char exits the fluid bed through the overflow pipe at 30 in. above the distributor plate. The vessel also contains an internal cyclone, which removes char fines from the exiting gas and returns them to the fluid bed.

The major components and stream flows of the complete PDU are illustrated in the simplified diagram in Figure 4. Feed is metered by rotary feeders from either of two lock hoppers to the fluid-bed carbonizer, from which product char overflows and falls into one of the two receivers, also lock hoppers. Residence time in the reactor is controlled by varying the solids feed rate.

Gas derived from carbonization of the feed is recycled through the system to fluidize the bed. Carbonization gases, along with recycle gas, leave the fluid bed, pass through the internal cyclone in the expanded section, and leave the vessel.

Figure 2. Carbonization PDU for clean coke process

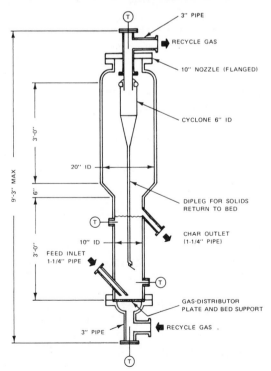

Figure 3. Fluid-bed carbonizer

The gas then passes through an external cyclone and into the gas-to-gas interchanger (shell and tube design). Hot carbonization effluent gas passes in downflow through the tube bundle, where it is partially cooled by heat exchange with clean recycle gas moving upflow on the shell side of the interchanger, and returning to the main gas heater. In the interchanger, the carbonization gas is also contacted with a spray of wash oil to remove tar mist and char dust which collect in the interchanger pot. The gas then passes to a water-cooled exchanger for final cooling to about 110°F (45°C), after which it passes in series through a wash-oil scrubber and a caustic scrubber for final cleanup of tar mist, char dust, and acid gases--particularly H_2S. The clean gas is then recycled by a compressor through the shell side of the interchanger to the main electric heater, where it is reheated to the temperature desired to maintain the fluid bed at design temperature. Actual temperature of gas from the heater varies with operating conditions in the reactor, but the temperature of gas exiting the heater is on the order of 100°F (55°C) higher than bed temperature. Net product gas is vented by a pressure regulator-controller through a wet-test meter and sampled for analysis.

The function of the water-injection system, Figure 4, is to maintain a concentration of about 8 vol % water vapor in the recycle gas while the gas is in contact with the hot alloy-metal surfaces of the main gas heater during second-stage carbonization. The presence of 8 vol % water vapor along with 50 ppm H_2S prevents formation of carbon deposits on the hot metal surfaces, which attain temperatures above about 1500°F during second-stage carbonization. Previous experience has shown that without the water vapor, carbon deposits grew large enough to impede gas flow through the heater significantly. Moreover, the carbon resulted in catastrophic carburization of the metal and destroyed the original recycle-gas heater. With water vapor and H_2S present, carbon formation is controlled, at least up to metal wall temperatures of 1550°F (845°C). Water is injected as a liquid into the interchanger shell-side gas inlet, where it is vaporized by external electric heaters. Most of the injected water condenses in the wash-oil quench in the interchanger pot and in the gas cooler. The remainder of the water condenses into the caustic-scrubber solution.

The wash-oil spray system is operated in different modes for the two stages of carbonization. For first-stage carbonization at 820°F, wash-oil quench of the gas occurs at the top of the interchanger to provide a washed-tube flow, which prevents plugging of the tubes by tar/char agglomerates. For second-stage carbonization at 1300°-1400°F (705°-760°C), wash-oil quench of the gas occurs below the interchanger tube bundle. In this mode, tube-exit gas temperature is controlled at about 900°F (480°C) to prevent condensation of tar within the tubes.

Fresh wash oil, at about 2 gal/hr, is metered continuously into the wash-oil scrubber to maintain a low concentration of tar and char fines in the oil system. Overflow from the level-controlled scrubber flows into the gas-cooler pot, from which the wash oil is pumped to the spray nozzle in the gas/gas interchanger. Wash-oil blowdown, including dissolved tar and suspended water and char, is removed from the interchanger pot. Wash-oil blowdown is screened to remove plus 50-mesh solids, heated to boil off contained water, and flash-distilled to separate heavy oil boiling above 540°F (280°C), which is used as part of the binder for coke production. The flash distillate is processed through a continuous distillation column at atmospheric pressure to separate a chemical oil distilling to 445°F (230°C) as overhead and fresh wash oil as bottoms. The wash oil consists essentially of mono- and dimethylnaphthalenes, acenaphthene, and quinoline bases.

The final system is caustic scrubbing, which controls H_2S concentration in the recycle gas. Normally, H_2S concentration in the gas is maintained in the range 50-100 ppm, which aids in controlling carbon deposition from the catalytic decomposition of the carbon components in the gas. Control of H_2S concentration may be effected by means of partial bypass of gas around the caustic scrubber and by varying the rate of metering fresh caustic solution (about 7% NaOH) into the scrubber. Generally, the PDU operators prefer to use the latter method.

Continuous Coal Oxidizer

An important adjunct to the carbonization PDU, the continuous coal oxidizer dries and preoxidizes the sized coal feed. The need for preoxidation of Illinois coal was discussed in the previous paper (3) on bench-scale studies, which showed that a mild surface oxidation of the coal diminishes its caking tendency sufficiently to prevent its agglomeration when it is heated at 800°F (425°C) in first-stage carbonization. Mild surface oxidation in this use refers to oxidation so slight that petrographic examination reveals virtually no change in the surface of the treated coal (4). Analyses of coal feed before and after mild oxidation at 350°F for 20 minutes indicate no significant change in the oxidized coal.

TABLE I.   Analysis of Oxidized Illinois Coal

Component		Analysis (wt %) [a]	
		Unoxidized	Oxidized
Carbon		75.1	75.5
Hydrogen		5.52	5.21
Oxygen [b]		11.6	12.0
Sulfur:	Total	1.69	1.74
	Inorganic	0.78	0.83
Ash		6.0	6.1

[a] Moisture-free basis.
[b] Determined by neutron activation analysis.
Includes oxygen in the ash.

In practice, sized coal feed is heated at 350°F (177°C)
for a 20-min residence time in a bed fluidized with air at
atmospheric pressure. The continuous coal oxidizer, shown in
Figure 5, is similar in design and operation to the carbonizer
reactor. The unit consists of a 10-in.-ID carbon-steel fluid-
bed reactor with a coal-feed and product-overflow system and an
electrically heated air supply. As in the PDU, the fluid bed
of the coal oxidizer is heated to and maintained at design
temperature by the heated fluidizing air. This unit is capable
of oxidizing up to 1 ton of coal per 24 hr of operation, and it
is normally operated at a coal feed rate of about 60 lb/hr.

Heated air is used in the existing coal oxidizer for
convenience only. In a larger operation, the oxidizer can be
operated with waste flue gas containing about 2% oxygen to
achieve adequate preoxidation of the feed coal.

Because the effluent gas from the coal oxidizer contains
only moisture and dust from the fluid bed, waste-gas cleanup is
accomplished by a small external cyclone and dust filters in
the vent system. The yield of dry, oxidized coal is essentially
a function of moisture content in the coal charged, which is
generally about 8% by weight.

Experimental Results From First-Stage Carbonization

Carbonization of the oxidized coal in the PDU is conducted
in two separate stages to avoid agglomeration of feed in the
fluid bed. Initial carbonization is effected at temperatures
in the range 800°-840°F (425°-450°C) to devolatilize the coal
partially and to produce a semichar, which can be fed subsequently

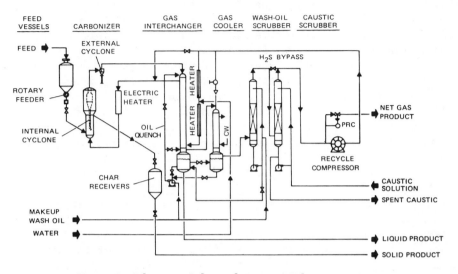

Figure 4. Schematic of clean coke process carbonization PDU

Figure 5. Continuous 10-in. coal oxidizer

into a fluid bed at 1400°F without agglomeration. The operating
limits on temperature for the first stage were determined by
bench-scale studies of the agglomeration problem. A nonagglomerating semichar was obtained at 800°F, and incipient
agglomeration was observed at about 850°F.

Although the primary function of first-stage carbonization
is to reduce the agglomerating property of the coal, about 70%
of the carbonization tar is produced in this stage. Gas production is low in the first stage, amounting to about 15 wt % of
the total gas produced. Analysis of the recycle gas shows its
composition (in mole %) to be methane, 71; ethane, 13; carbon
monoxide, 10; hydrogen, 2; and C_2-to-C_4 hydrocarbons, 4.

Tests were made in the PDU to study the effect upon volatile-
matter and sulfur contents in the semichar of bed temperature,
residence time, system pressure, and H_2S concentration in the
recycle gas. The variation in volatile-matter content with
temperature and residence time is illustrated in Figure 6,
which shows that volatile-matter content varies inversely with
temperature and residence time and is independent of pressure
over the range 15-160 psia. In the figure, a few data points
from a 1-in. continuous bench-scale unit were included to
extend the pressure range to 15 psia. Generally, about half
the volatiles were eliminated from the coal in first-stage
carbonization, and all the semichar products were processed
through second-stage carbonization without agglomeration.

Response of sulfur content to residence time and temperature
of first-stage carbonization is shown graphically in Figure 7.
Sulfur content of the semichar product also varies inversely
with temperature and residence time and is independent of
pressure over the ranges studied; conditions included temperatures of 800° and 840°F, residence times of 20-80 min, and
pressures of 80-160 psia. In these tests, H_2S concentration
was controlled in the range 50-100 ppm. However, a test was
run at 800°F, 52 min residence time, 120 psia, and H_2S concentrations varying from 300 to 2000 ppm. The semichar product from
this test contained 1.77% sulfur, which is in the range normally
attained with low H_2S concentration in the recycle gas. Because
gas evolved from devolatilization of the coal was recycled in
the system to fluidize the bed, gas composition was not a
controllable parameter, except for concentration of acid gases.

Temperature and residence time thus appear to be the only
variables having an effect on volatiles and sulfur remaining in
the semichar product. The data in Figures 6 and 7 indicate
that, at the temperatures deemed feasible, first-stage devola-
tilization/desulfurization is adequate in about 20 min residence

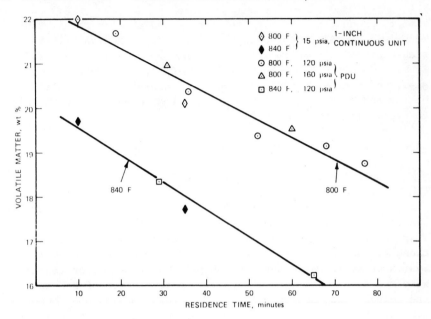

Figure 6.   Effect of temperature and residence time on coal devolatilization

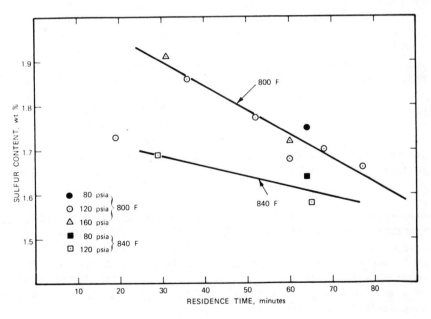

Figure 7.   Effect of temperature and residence time on semichar sulfur content

time to produce a suitable feed for second-stage carbonization.
To provide a consistent data base for process design, first-
stage carbonizations in the PDU are now routinely run at 820°F,
25 min residence time, 165 psia, and 100 ppm H_2S in the recycle
gas. Under these conditions, coal containing 35% volatile
matter and about 2.2% sulfur is converted to semichar containing
20% volatile matter and 1.7% sulfur.

Experimental Results From Second-Stage Carbonization

Completion of devolatization and desulfurization required
the higher reaction temperatures of the second-stage carbonization. Tests were made in the PDU to study effects of temperature
and residence time in the fluid bed upon sulfur content of the
char produced in second-stage carbonization. The ranges for
these variables were 1250°-1450°F (675°-790°C) and 40-200 min.
System pressure was generally 165 psia, although other tests
were run at 80, 120, and 150 psia. Concentration of H_2S in
the recycle gas was controlled in the range 50-100 ppm, but
concentrations as high as 1500 ppm H_2S were studied in special
tests.

Results from the studies of temperature and residence-time
effects are shown graphically in Figure 8. For clarity only
data obtained at 1250°, 1325°, and 1400°F are used. Sulfur
content of the semichar feed for the test series ranged from
1.65-1.80% and averaged about 1.70%. The quantity of char
produced at a single set of conditions was usually about 600 lb,
and more than 1600 lb of char was produced during the longest
test. Sulfur content of the char products ranged from 1.1% at
the mildest condition to 0.2% at 1400°F and 190 min residence
time.

Temperature exerted the greatest effect on desulfurization
of the char, as indicated in Figure 8; a temperature increase
of 75°F provided a lower char sulfur content than increasing
residence time threefold or even fourfold. For example, 40 min
residence at 1400°F was the equivalent of 157 min at 1325°F in
producing char containing 0.6% sulfur. It is apparent from the
data that desulfurization of the feed occurred rapidly during
the initial period of heating and devolatilization. Chars
having sulfur contents in the range 0.7-0.8% were obtained in
40-50 min residence time at fluid-bed temperatures greater than
about 1300°F. Desulfurization below about 0.8% sulfur proceeded
at a much slower rate, and the rate then appeared to be almost
completely linear with time.

All the data in Figure 8 were obtained at a system pressure of 165 psia, except for the one point, indicated on the 1325°F line, which was obtained at 100 psia. This point shows that the effect of pressure over the range of 100-165 psia is not discernible in the sulfur content of the char from the continuous fluid-bed reactor. Other tests in the PDU gave similar results; for example, at 1370°F and about 90 min residence time, the sulfur contents of the char products were 0.67 and 0.66% at 120 and 150 psia, respectively.

The inability to demonstrate a pressure effect in the continuous reactor of the PDU results from the interdependence of variables in the system. For example, intentionally changing the pressure automatically changes the mole fraction of hydrogen in the gas, which changes the hydrogen-to-sulfur ratio in the fluid bed. In addition, a smaller mass of lower-pressure (lower density) gas must be recirculated through the bed to maintain the same degree of fluidization, and lower-pressure gas must be hotter to effect the desired heat transfer to maintain bed temperature. Because of this, char particles near the bottom of the bed are contacting hotter gas and are momentarily heated to temperatures greater than the average bed temperature. These competing forces combine to mask the pressure effect in the continuous reactor. However, the effect of increased pressure helps to decrease the sulfur content of the product char. This has been demonstrated by numerous investigations and by our bench-scale tests in batch reactors (3), which were suited for studying the pressure effect separately from the other variables.

An important dependent variable, for which a continuous reactor is best suited, is average sulfur content of the bed. In a batch unit, the sulfur content of the bed declines throughout the reaction time, but in a continuous unit, the average sulfur content of the fluid bed remains essentially constant, because of the continuous addition of sulfur with the feed and depends upon the sulfur content of the feed and upon the feed rate. This factor contributes significantly to the observed benefit of very long residence times, which were achieved in the PDU by greatly reduced feed rates. The consequent reduction in rate of sulfur addition to the bed resulted merely in a lower average sulfur content in the bed. From these considerations, it may be concluded that the important variables in desulfurization of char in the PDU are temperature and average sulfur content of the bed, provided there is a significant concentration of hydrogen in the fluidizing gas.

Composition of the fluidizing gas was not a controllable variable, except for H_2S concentration, because process-derived gas was recycled in the system. The gas was composed almost entirely of methane, hydrogen, and carbon monoxide. Carbon dioxide was present only in tenths of a percent because of caustic scrubbing to control H_2S concentration. At 165 psia, hydrogen concentration in mole percent varied from the low 20s at 1250°F to the low 40s at 1400°F; methane concentration ranged from the low 70s to the mid 50s; carbon monoxide concentration was nearly independent of reaction temperature but varied from a high of about 6% at the shorter residence times to about 2% at the longer times. Hydrogen concentration was also sensitive to residence time and increased 3-5 percentage points between the shortest and longest times.

The effect of several H_2S concentrations in the recycle gas was studied in tests conducted at 1400°F (Table II). The semichar feed for these tests contained about 1.75% sulfur. The tests were run primarily to generate data relevant to the design of larger fluid-bed reactors, in which bed height would be significantly greater than the 30-in. bed height in the existing PDU. Concentration of H_2S increases in the fluidizing gas as it passes up through the bed, and at bed depths envisioned for reactors designed for 100 tons or more of feed per day, average H_2S concentration within the fluid bed might easily reach 1000 ppm.

TABLE II. Effect of H_2S in Fluidizing Gas on Char
          Sulfur[a]

H_2S Concentration (ppm)	Char Product (wt % S)
50-100	0.21
500	0.71
1000	0.69

[a] 1400°F, 165 psia, 190 min residence time.

The data in Table II show a significant deterioration in char sulfur content, from 0.2% to 0.7%, when the H_2S level in the gas entering the fluid bed was increased from the normal 50-100 ppm to the 500-ppm level. Interestingly, increasing the H_2S level to 1000 ppm did not have any perceptible additional effect. These tests were limited by the exposure of all fluidized particles to the high H_2S concentration in the incoming gas, whereas all particles in a deeper fluid bed would also be exposed about half of the time to gas of much lower H_2S concentration. In a more recent test, a low-sulfur char was fed at 1400°F to a bed fluidized with gas containing 1200-1500 ppm H_2S. In this case, the sulfur content of the char increased from 0.4% to 0.6% during a 104-min residence time in the fluid bed. This resulfidation of the char was expected, but its modest impact on the product quality was gratifying. About three fourths of the pickup was reported as inorganic sulfur, and the balance was organic sulfur.

The design of the PDU does not lend itself to more definitive studies of deep fluid beds; however, the results do suggest that desulfurization will be somewhat less efficient in commercial-size carbonizers. Nevertheless, commercial production of char having an acceptable sulfur content appears feasible.

To indicate the reproducibility of data points, Table III presents analyses of the consecutive receivers of char produced during 10-day runs at 1400°F, 165 psia, and residence times of 46 and 190 min. The total quantities of feed for the tests were 2170 and 875 lb, respectively. The data show excellent reproducibility, considering that observed variations in the products are the cumulative effects of variations in the semichar feed and process conditions, plus the repeatability of sampling and analysis.

TABLE III.   Uniformity of Product Char from Carbonization PDU[a]

	Char Sulfur Content (wt %)	
Receiver	46 min	190 min
1	0.69	0.26
2	0.70	0.21
3	0.68	0.18
4	0.61	0.18
5	0.68	0.18
6	0.66	0.25
7	0.68	--
8	0.71	--
Average	0.68	0.21

[a] 1400°F, 165 psia.

Of interest also are the incremental changes in concentration of the various forms of sulfur originally present in the raw coal. These changes are illustrated by the analytical

results in Table IV for the feed and products involved in the
Clean Coke Process. Forms of sulfur are shown simply as organic
and inorganic, because the inorganic sulfur was nearly all
pyritic and contained at most 0.05% sulfate sulfur. Coals from
several mines in central Illinois were evaluated, and all
samples were quite similar. However, only coal from No. 24
mine of the Old Ben Coal Co. was processed in the carbonization
PDU, and the data in Table IV were obtained with this coal.

TABLE IV. Incremental Change in Forms of Sulfur

Processing Step	Sulfur Forms (wt %)	
	Organic	Inorganic
As-mined coal	1.00	2.50[a]
Cleaned and sized carbonization feed	1.13	1.00
Semichar (first-stage product)	0.91	0.81
Char (second-stage product, 1400°F)	0.13	0.08

[a] Inorganic sulfur is pyritic sulfur plus about
0.05% sulfate sulfur in the as-mined coal.

Typically, the run-of-mine coal sample contained about 1%
organic sulfur and 2% or so inorganic sulfur. Conventional wet
cleaning of this coal by gravity separation and tabling operations removed inorganic sulfur selectively and produced a clean
coal feed which contained about 1% inorganic and 1.1% organic
sulfur. Devolatilization and desulfurization in first-stage
carbonization at 820°F removed about 33% of the sulfur in each
form, and the ratio of organic to inorganic remained about the
same in the semichar product as in the feed coal. (The coal-
drying and preoxidation treatment did not affect either the
amount or the forms of sulfur.) Final devolatilization and
desulfurization in second-stage carbonization at 1400°F removed
about 90% of each form of sulfur in the semichar feed, and the
lowest sulfur char contained only slightly more organic than
inorganic sulfur.

Both forms of sulfur contributed substantially to the
production of H_2S during carbonization. This fact is illustrated
by the data in Table V, which show the quantity of sulfur in

each form that was converted and the quantity recovered as H_2S in the caustic scrubbing solution. Converted pyritic sulfur contributed about two thirds of the H_2S produced by first-stage carbonization at 820°F, and the inorganic sulfur remaining in the semichar contributed about half of the H2S produced during second-stage carbonization at 1400°F. Assuming that all the reacting inorganic sulfur is converted to H_2S, Table V data show that about 70% of the organic sulfur reacting at 820°F was converted to H_2S, and about 95% was converted to H2S at 1400°F. Thus, reaction conditions in both stages are adequate to convert both types of sulfur compounds to easily recoverable H2S. The remainder of the sulfur liberated from the coal during pyrolysis is recovered as organic compounds in the liquid products, which contain 1.1-1.3% sulfur.

TABLE V. Conversion of Forms of Sulfur to H2S[a]

| | Distribution of Sulfur (lb) | | | |
	Organic	Inorganic	Total	Recovered as H2S
Coal feed	9.5	11.3	20.8	--
Converted sulfur (at 820°F)	3.0	4.4	7.4	6.5
Semichar feed	6.5	6.9	13.4	--
Converted sulfur (at 1400°F)	4.6	4.4	9.0	8.5
Char product	1.9	2.5	4.4	--

[a] Basis: 1000 lb coal feed.

Material Flows and Mass Balance

Material flows and mass balance through the two stages of carbonization of Illinois coal are shown in the simplified flow diagram of Figure 9. The products of first-stage carbonization of the coal are (in wt %) semichar, 83.0; tar, 11.9; water, 2.5; fuel gas, 1.1; and acid gases, 1.5. At 1400°F and 77 min residence time, the products of second-stage carbonization of the semichar are (in wt % of the coal) low-sulfur char, 62.9; tar, 5.0; water, 2.0; process gas, 6.3; and acid gases, 6.8.

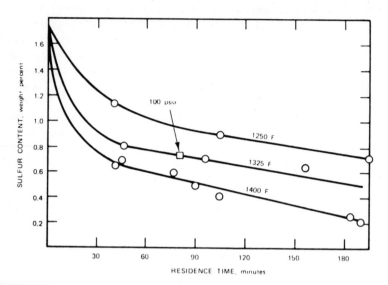

Figure 8. *Effect of temperature and residence time on char sulfur content (165 psia)*

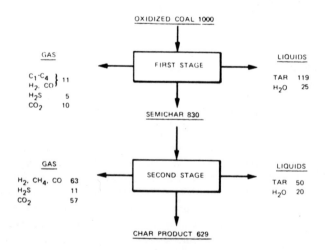

Figure 9. *Material flow and weight distribution during staged carbonization*

Weight balances from PDU tests were in the range 99-102%
of the weights fed into the PDU. Special material-balance
tests were run to establish elemental balances on carbon and
sulfur, Table VI. Overall weight balance in these tests accounted
for 99.3% of the weight input. The carbon-balance data show an
overall accountability of 99.7%, and the sulfur-balance data
show an overall accountability of 98.5%.

TABLE VI.   Elemental Balance - Overall[a]

	Input (lb)			Output (lb)	
	C	S		C	S
Coal	724	20.7	Char	521	2.8
Wash Oil	1981	25.3	Wash Oil & Tar	2099	27.1
			Gas	76	15.4
Total	2705	46.0	Total	2696	45.3

[a]Basis: 1000 lb Coal Feed.

Conclusions

The carbonization PDU of the Clean Coke Process has proved
to be a valuable research tool in demonstrating, on a substantial
scale, the efficient desulfurization of Illinois coal in a continuous, pressurized fluid-bed carbonizer. After conventional
cleaning and sizing, coal was processed during sustained operation of up to 10 days through two separate stages of carbonization, which together removed more than 90% of the coal sulfur
and produced char containing as little as 0.2% sulfur. First-
stage carbonization at about 820°F served primarily to produce
a nonagglomerating semichar feed for high-temperature carbonization, but the first-stage carbonization removed about 33% of
the sulfur from the coal and produced about 66% of the total
tar. Second-stage carbonization removed up to 90% of the
remaining sulfur and produced low-sulfur char, hydrogen-rich
fuel gas, and tar. Both organic and inorganic sulfur were
removed with equal facility by the carbonization process.

In addition to the data on desulfurization and process
yields and chemistry, the PDU has provided much useful engineering
information that was needed for the design of a 100-ton/day
pilot plant. The pilot-plant process-design work is currently
in progress, and continued testing in the PDU will generate
data needed for the design. Concurrently, studies are in
progress on carbonization of high-sulfur coals from other major

seams of national interest.   Coal from the Kentucky No. 9 seam
is being processed at present, and this will be followed by
testing of a Pittsburgh seam coal.

Acknowledgments

The information contained in this paper and generated in
the carbonization PDU would not have been possible without the
design modifications and engineering assistance of George A. Ryder
and John Stipanovich of the Chemical Engineering Design group.
Their contributions to the success of the PDU program were
invaluable and are sincerely appreciated.

Literature Cited

1.   Schowalter, K. A., Boodman, N. S., "Clean Coke Process,"
     presented at the Symposium on Coal-Conversion Processes,
     Meeting of the American Institute of Chemical Engineers,
     Philadelphia, November 1973.
2.   Schowalter, K. A., Petras, E. F., "Aromatics From Coal,"
     presented at the Symposium on Aromatics, Meeting of the
     American Institute of Chemical Engineers, Houston, March
     1975.
3.   Johnson, T.F., Krupinski, K. C., Osterholm, R. J., "Clean
     Coke Process: Fluid-Bed Carbonization of Illinois Coal,"
     presented at the Symposium on New Cokemaking Processes
     Minimizing Pollution, American Chemical Society Meeting,
     Chicago, August 1975.
4.   Gray, R. J., Krupinski, K. C., "Use of Microscopic Procedures
     to Determine the Extent of Destruction of Agglomerating
     Properties in Coal," presented at the Coal Agglomeration
     and Conversion Symposium, West Virginia University,
     Morgantown, May 1975.

Support received from the U.S. Energy Research and Development
Administration under Contract No. E(49-18)-1220.

Hydrodesulfurization of Coals

DONALD K. FLEMING, ROBERT D. SMITH, and MARIA ROSARIO Y. AQUINO

Institute of Gas Technology, 3424 South State St., IIT Center, Chicago, IL 60616

The Institute of Gas Technology (IGT) is engaged in a program to develop a patented process (1) to desulfurize coal to produce a clean, solid fossil fuel product. This process — called the IGT Flash Desulfurization Process — uses a combination of chemical and thermal treatment of the coal. The present effort, funded by the U.S. Environmental Protection Agency (EPA Contract No. 68-02-2126), aims to determine the operating parameters of the primary reactors of the system. The objective of the process is to be able to treat any coal so that the resulting solid fuel can be directly consumed in an environmentally satisfactory manner.

Equipment scaled to laboratory, bench, and continuous process development unit (PDU) investigations is being used in the project. Each coal tested is subjected to preselected conditions of temperature, heating rates, and residence time in a reducing atmosphere. After treatment, the material is chemically analyzed to determine the degree of sulfur removal. Results from tests with four different high-sulfur coals (all from abundant Eastern U.S. seams) show good sulfur reduction; calculated sulfur dioxide emissions of the treated material are below the present Federal EPA standards of 1.2 lb/10^6 Btu for direct combustion of the solid fossil fuel product.

Coals Tested

Several candidate coals were characterized for sulfur content, seam location, and quantity available. Subbituminous and lignite coals were eliminated because of their low content of naturally occurring sulfur. Four bituminous coals selected for testing were the following:

1. Western Kentucky No. 9, containing 3.74% sulfur (run-of-the-mine)
2. Pittsburgh Seam (from a West Virginia mine), containing 2.77% sulfur (highly caking)
3. Pittsburgh Seam (from a Pennsylvania mine), containing 1.35% sulfur (high ash content)

4. Illinois No. 6, containing 2.43% sulfur (washed).
These coals were selected without regard for their relative content of pyritic and organic sulfur, because a universal coal
desulfurization process should be able to minimize any type of
sulfur in the coal.

Pretreatment

The coals selected are all caking coals; an oxidative pretreatment, prior to hydrodesulfurization, is included in the
system to eliminate this caking. Similar to the pretreatment used
in gasification processes, this system would operate with air-
fluidization at near-atmospheric pressure using coal feedstock at
—14 mesh. Pretreatment tests were conducted in a batch reactor to
determine the proper pretreatment conditions for each coal.
Ranges of temperatures, relative-air rates, fluidization velocities, and residence times were tested. Results of these tests
indicated that a temperature of 750°F and a gas velocity of 1 ft/
sec were necessary to achieve satisfactory pretreatment. ("Satisfactory pretreatment" is defined, for purposes of this program, as
pretreatment that conditions the coal sufficiently to pass an
empirical caking test.) For the Western Kentucky No. 9 coal, a
residence time of 30 min and an oxygen consumption of 1 SCF/lb of
coal were adequate conditions in satisfying pretreatment; other
coals required different degrees of pretreatment. Based on test
results, parameters of residence time and oxygen consumption were
adjusted for each feedstock to yield a noncaking material.
Approximately 8% - 12% of the coal is consumed during pretreatment, generating steam for the rest of the system and a low-
Btu off-gas that can be burned onsite to provide process steam or
to generate power. Approximately 25% - 30% of the coal sulfur is
removed during pretreatment; this sulfur becomes primarily SO_2 in
the low-Btu pretreatment off-gas.
Pretreatment not only prevents caking, it also increases the
sulfur removal in the subsequent hydrotreating step. Figure 1
represents two series of thermobalance tests made with Western
Kentucky No. 9 coal. One test series was made with crushed and
screened coal; the other test used crushed, screened, and pretreated coal as feed for hydrodesulfurization. The results show
that the 70% sulfur removal (based on the initial feed), as
achieved with untreated coal feed, was increased to 95% by using a
pretreated feed. Sulfur removed initially during pretreatment,
primarily pyritic sulfur, is also readily attacked in low-
temperature hydrotreatment; the improved sulfur removal after pretreatment is obtained by more complete removal of organic-type
sulfur. The more complete removal may be caused by the increase
in coal pore size during pretreatment, resulting in better contact
of the hydrogen with the coal sulfur and larger passages for
removal of the resultant hydrogen sulfide. Also, oxidative treatment of sulfur compounds is expected to activate the sulfur for

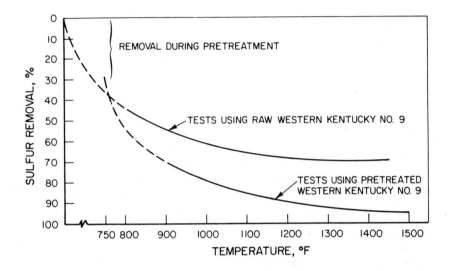

Figure 1. Sulfur removal indicating effect of coal pretreatment

further reactions.

Hydrodesulfurization Results

Thermobalance Tests. A preliminary evaluation of the desulfurization of each coal was performed in a thermobalance, a
laboratory device that can weigh continuously a sample exposed to
a controlled environment of temperature, pressure, and contacting
gas composition. (A complete discussion of experimental equipment, operating procedures, and laboratory analytical technique is
presented in the final report (2) prepared under previous EPA
contract No. 68-02-1366.) Results are used to determine sulfur
removal from the feedstock under a wide range of conditions. A
total of 122 thermobalance tests have been performed in this
program.

Samples for thermobalance tests were prepared with +40 mesh
pretreated coal. This feed was placed in the sample basket and
lowered into the heated zone. At high gas flow rates, relative
to the amount of coal, the thermobalance operates as a differential reactor. Heating rates of $5^{\circ} - 20^{\circ}F/min$ were used, up to
terminal temperatures of $1000^{\circ} - 1500^{\circ}F$. Soaking times at the
terminal temperature were varied from 0 to 5.5 hr. The laboratory
analyses of the treated coal provided sulfur content by types,
including pyritic, sulfide, sulfate, and organic sulfur. Because
samples were small, more complete characterization could not be
achieved.

Figures 2 - 5 show sulfur removal achieved relative to the
initial coal, as obtained from the thermobalance tests, for each
of the four specified coals. For these tests, the reactant gas
was pure hydrogen. Heat-up rates were $5^{\circ} - 20^{\circ}F/min$, and soaking
times were 30 to 60 min. The pressure was near atmospheric.
Results of tests with the Western Kentucky No. 9 coal are shown in
Figure 2. An apparent increase in pyritic sulfur content during
pretreatment may be caused by elutriation in the fluidized-bed
pretreater; heavier pyrites are retained in the residue from pretreatment. Most of the reduction in organic sulfur content is
attributable to devolatilization. The figures show that nearly
all of the pyritic sulfur was removed and that 88% of the organic
sulfur was removed at $1500^{\circ}F$. Sufficient total sulfur removal has
been achieved in all the tests above $1400^{\circ}F$ so that SO_2 emissions
from combustion of the treated product would be below 1.2 $1b/10^6$
Btu.

The sulfur reduction obtained for the Pittsburgh Seam coal
(from a West Virginia mine) is shown in Figure 3. At all temperatures above $1300^{\circ}F$, 97% of the pyritic sulfur was removed.
Organic sulfur content was reduced by 85% at $1500^{\circ}F$. The reduction in total sulfur content exceeded 90% at $1500^{\circ}F$. The coal
treated at any temperature above $1400^{\circ}F$ could be burned in compliance with present SO_2 emission limitations.

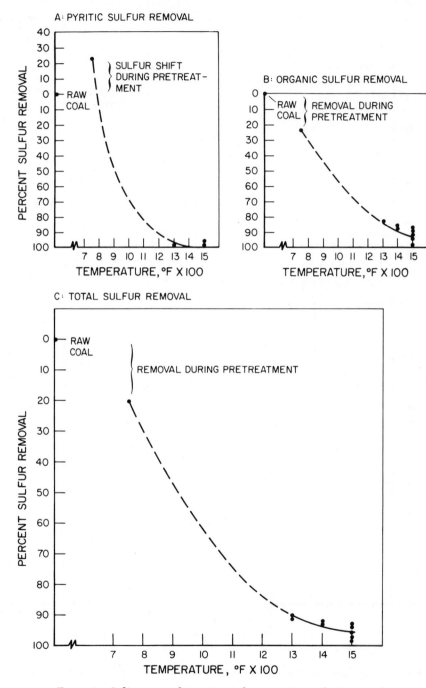

Figure 2. Sulfur removal in pretreated western Kentucky No. 9 coal

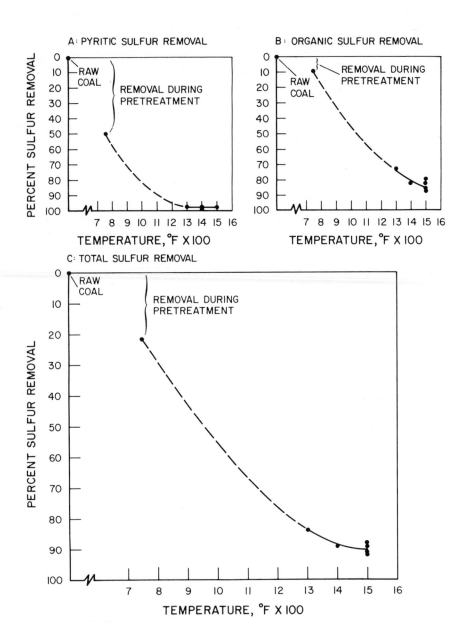

Figure 3.   Sulfur removal in pretreated Pittsburgh seam, West Virginia mine, coal

Results for the Pittsburgh Seam coal (from a Pennsylvania mine) are shown in Figure 4. Pyritic sulfur content has been reduced 95% at 1300°F and reaches 100% at 1500°F. Only 60% - 80% of the organic sulfur was removed in this temperature range; however, this coal has a lower initial sulfur content. All the resultant treated material, therefore, yields low SO_2 emission values.

The removal of sulfur from Illinois No. 6 coal is shown in Figure 5. Pyritic sulfur has been completely removed at 1300°F and above. Organic sulfur content has been reduced by 80% and total sulfur content by more than 90% at 1500°F. Again, enough sulfur is removed so that all samples meet emission requirements.

The results for all four coals are superimposed in Figure 6; this direct comparison indicates that all of the coals behaved similarly.

In the tests described above, each sample was heated in the thermobalance slowly (5° - 20°F/min) to its terminal temperature. A series of thermobalance tests with Western Kentucky No. 9 coal used rapid heating. Rapid heating is accomplished by first heating the reaction zone of the thermobalance to the desired temperature, then lowering the sample basket into the preheated zone. Most of the total weight change occurs in the first few seconds that the sample is in the hot zone. After 15 min, the weight changes only slightly, regardless of the ultimate residence time. The total of sulfur removed, however, increases with residence time when rapid heating is used. Reduction of sulfur content by 95% has been achieved in 2-hr residence time at 1500°F; however, samples subjected to 60 min or more met the EPA emission limits for SO_2.

Batch Reactor Tests. A batch reactor has been used with the Western Kentucky No. 9 and Illinois No. 6 coals to substantiate the results of the thermobalance tests and to extend testing to other development phases. The batch reactor is constructed from 1-1/2 in. pipe with a sintered-metal plate for gas distribution. This reactor operates in a fluidized-bed mode, similar to the anticipated operation of the full-scale plant, and can be subjected to programmed heating rates with external heaters controlled by internal bed thermocouples. The batch reactor can treat larger samples, and the treated product is more completely characterized analytically. The batch reactor has been operated with hydrogen gas at near-atmospheric pressure; coal (or pretreated coal) feedstock was screened at —20+80 mesh. A total of 128 batch reactor tests (including pretreatment evaluations) have now been performed.

Batch tests, with conditions similar to those in the thermobalance experiments, were made at terminal temperatures of 1400° and 1500°F. Results were excellent, with as much as 98.6% of the sulfur content removed at 1500°F; these results agree well with the thermobalance tests. The treated material would produce SO_2 emissions well below the EPA limitations, confirming earlier thermobalance results.

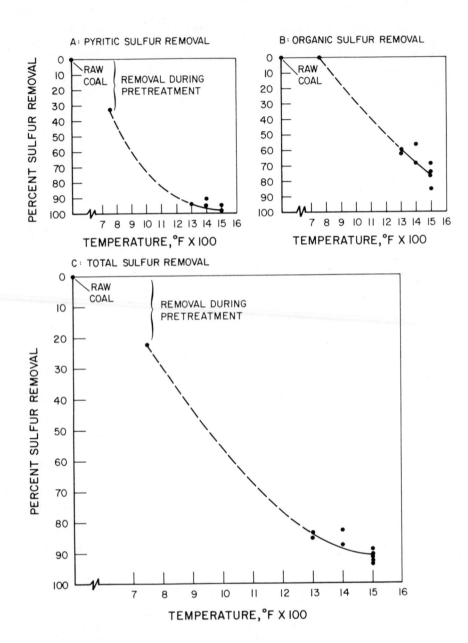

Figure 4. Sulfur removal in pretreated Pittsburgh seam, Pennsylvania mine, coal

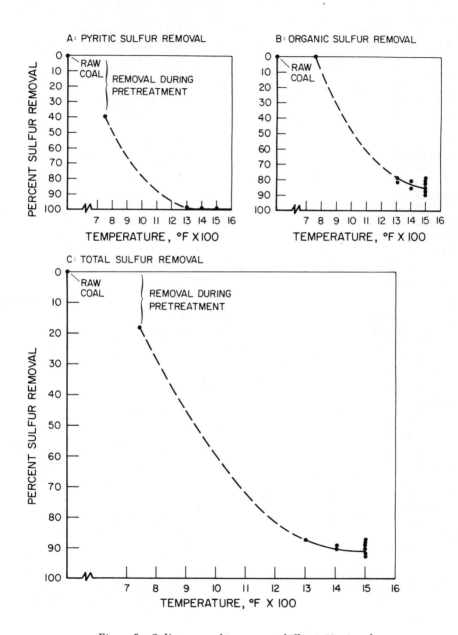

Figure 5. Sulfur removal in pretreated Illinois No. 6 coal

Table I presents typical results from two batch reactor tests. (Detailed data will be presented in the final report under EPA Contract No. 68-02-2126 (3).) For these tests, the product recovery is about 60%; the remainder of the coal has been gasified (and pretreated) into low-Btu gas that can be upgraded to pipeline quality or consumed onsite. The heating value of the treated product is about 5% less than that of the feedstock, primarily because of both the lost heat content associated with burning the coal-sulfur and the increased ash content of the product. The quantity of volatile matter contained in the treated product has been reduced significantly; modified combustion equipment may be required to consume the desulfurized coal. Alternatively, as in another IGT patent, the treated product can be recombined with the hydrocarbons produced during the treatment (after oil hydrodesulfurization) to improve the combustion characteristics.

Figure 7 presents the batch reactor results for tests on Western Kentucky No. 9 coal. At any treatment temperature, the sulfur removal from the coal approaches a maximum asymptotically with soaking time. Based on these tests, a reasonable design condition would be 90 min of reaction time at $1500^\circ F$.

PDU-Scale Tests

Work has now progressed to larger equipment, using a 10-in. fluidized-bed unit that can be fed continuously at rates between 25 and 100 lb/hr. The 10-in. unit has been used to verify pretreatment operating conditions on a continuous basis. Pretreated feedstock has been prepared on this unit for hydrodesulfurization runs that are (at the time of preparation of this paper) now planned.

This unit will be used to collect data for material and energy balances, stream characteristics, economics, and design specifications for a larger installation.

Preliminary engineering evaluation of the system, based on the in-house data base at IGT, indicates that the process should have favorable economics. Sufficient hydrogen should be evolved from the coal during devolatilization that a supplemental hydrogen source will not be required; rather, the hydrotreatment gas may be recycled, after hydrogen sulfide removal. Further, the quantity of low-Btu off-gas generated during the hydrotreatment will result in a substantial bleed stream from the recycled gas; the value of this bleed stream is significant and will be instrumental in reducing the cost of the treated solid product to be competitive with other clean fuels. Confirmation of these preliminary evaluations, however, must await data from PDU and, eventually, pilot plant tests.

Table I. Typical Batch Reactor Runs with Specified Feedstock

	BR-76-3		BR-76-34	
Run no.	Run-of-Mine W. Ky. No. 9		Washed Illinois No. 6	
Coal type	Feed	Product	Feed	Product
Sample				
Laboratory ID no.	26498	31996	33293	34428
Terminal temperature (°F)		1500		1500
Heat-up rate (°F/min)		5		5
Soaking time (min)		30		30
Proximate analysis (wt %)(as received)				
moisture	5.8	0.8	2.4	0.4
volatile matter	36.3	3.3	34.0	3.3
ash	10.6	18.3	8.1	12.5
fixed carbon	47.3	77.6	55.5	83.8
total	100.0	100.0	100.0	100.0
Ultimate analysis (wt %)(dry basis)				
ash	11.24	18.43	8.31	12.51
carbon	70.00	78.70	73.90	83.40
hydrogen	4.54	0.95	4.81	1.06
sulfur				
sulfide	0.02	0.15	0.01	0.05
sulfate	0.64	0.00	0.13	0.05
pyritic	1.13	0.02	0.82	0.03
organic	1.95	0.46	1.47	0.49
total	3.74	0.63	2.43	0.62
nitrogen	1.53	0.78	1.60	0.90
oxygen	8.95	0.51	8.95	1.51
total	100.00	100.00	100.00	100.00
Heating value (Btu/lb)	12,454	11,967	13,168	12,793
Solids recovery (%)		62.62		62.70
Total sulfur removal (%)		89.45		84.00
Pyritic and organic sulfur removal (%)		91.96		86.58

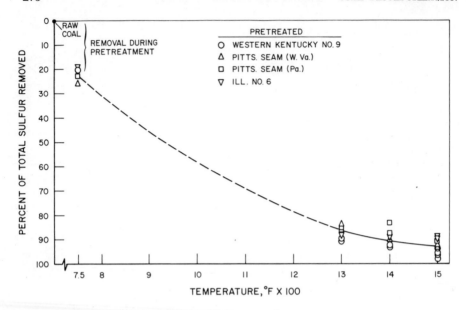

Figure 6.   *Percent sulfur removed vs. temperature*

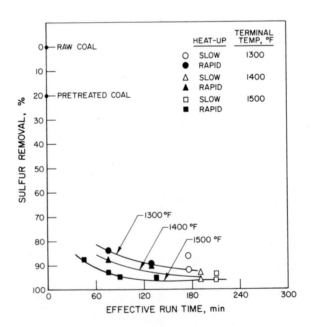

Figure 7.   *Degree of sulfur removal in batch reactor tests
as a function of effective run time (total time during heat-
up or soaking at temperature greater than 750°F)*

Conclusions

Laboratory- and bench-scale data indicate that acceptable hydrodesulfurization of coals can be achieved with the IGT Flash Desulfurization Process. Pretreatment of the coal enhances the removal of sulfur in the subsequent hydrodesulfurization step to produce a solid fuel that can be burned in conformance with the present Federal EPA New Source Performance Standards of 1.2 lb $SO_2/10^6$ Btu. Work is progressing to prove the concept on larger continuous, PDU-sized equipment. The complete flow sheet for the process has not yet been defined; therefore, economic factors are unknown.

Acknowledgement

The work in this coal desulfurization program has been performed under contract to the U.S. Environmental Protection Agency, Research Triangle Park, North Carolina, and EPA permission to publish these results is gratefully acknowledged.

Abstract

The IGT process for producing low-sulfur, solid fossil fuel from coal is based upon chemical and thermal treatment. Sulfur is removed from the coal by hydrogen treatment at elevated temperature and near-atmospheric pressure. Several coals, from abundant Eastern U.S. seams, have been tested in both laboratory- and bench-scale hardware. An oxidative pretreatment of the coal both prevents coal from caking and improves the overall sulfur removal. The data indicate that enough sulfur, both pyritic and organic types, is removed so that the treated product can be burned directly in conformance with present Federal EPA NSPS for SO_2 emissions. The program, sponsored by EPA, is now testing the concepts in a larger, continuous fluidized-bed unit to verify the bench-scale results.

Literature Cited

1. Lee, B. S. and Schora, F. C., "Desulfurization of Coal," U.S. Patent 3,640,016 (1972) February 8.
2. Institute of Gas Technology, "Development Program for Treatment of Coal to Produce Low Sulfur, Solid Fossil Fuel," IGT Project 8945, EPA Contract No. 68-02-1366, Chicago, 1977. (To be published.)
3. Institute of Gas Technology, "Pilot Plant Study of Conversion of Coal to Low-Sulfur Fuel," IGT Project 8973, EPA Contract No. 68-02-2126, Chicago, 1977. (Program still in progress.)

21

Improved Hydrodesulfurization of Coal Char by Acid Leach

ANN B. TIPTON

Occidental Research Corporation, La Verne, CA 91750

Occidental Research Corporation's (ORC) Flash Pyrolysis coal liquefaction process (1) is designed to handle caking coals in a single-stage reactor without pretreatment. The process produces a maximum liquid yield and char co-product. When high-sulfur, bituminous coals are used as feed for the process, the char, like the coal, is also high in sulfur. The value of this char as a feed for utility boilers would be increased greatly if its sulfur content were lowered to meet the EPA sulfur emission standard (1.2 lb SO_2/MM Btu). Hence, concurrent with the development of the Flash Pyrolysis coal liquefaction process, char desulfurization research has been performed. From this research, a two-step acid leach/hydrodesulfurization (AL/HDS) process has been developed (2). Acid leach, the first step of the process, removes primarily iron and calcium sulfides from the char. The absence of these minerals improves significantly the hydrodesulfurization step by allowing operation at shorter residence times and at higher H_2S concentrations.

The original research for the acid leach/hydrodesulfurization process, reported by Robinson (2), studied the effects of temperature, pressure, residence times, superficial velocity, particle size, char preparation method, coal class (bituminous and subbituminous), acid type, and H_2S concentration in the treatment gas (H_2). In most of this work, chars prepared by pyrolysis of coals in a static bed under an inert gas blanket were used. However, Robinson established that acid-leached Flash Pyrolysis chars have faster hydrodesulfurization rates than acid-leached static bed chars. Therefore, the work reported here was performed to establish the data base required for the design of an AL/HDS pilot plant for Flash Pyrolysis char from high-sulfur, bituminous coal.

Experimental

The char used in this study was made in ORC's Flash Pyrolysis coal liquefaction pilot plant by a single pass of -200 mesh Western Kentucky No. 9 coal through the pyrolysis reactor at 1200°F for 2.5 sec. The following additional process steps were performed in laboratory batch reactors. The single-pass Flash Pyrolysis char was oxidized at 1600°F for 0.1 sec and then held in a fluidized bed under nitrogen at 1600°F for 1 hr. The oxidation represents a char burning step in the coal liquefaction process and the additional heating simulates continuous char recycle.

The Flash Pyrolysis char was acid-leached five times with 6N hydrochloric acid at 80°C for 5 min with a 2:1 acid-to-char ratio. Residual acid was removed by water rinsing.

The hydrodesulfurization reactor and chemical and physical property analyses were described in detail by Robinson (2). The hydrodesulfurization reactor was built around a 1.2-m x 19-mm i.d. quartz cylinder in which 10-g samples were treated at 1600°F, 65 psia, and 0.15 ft/sec superficial velocity. Before the treatment gas was introduced, the reactor and sample were heated to temperature and held for 15 min with nitrogen as the fluidization medium. The system was also purged with nitrogen during cool down.

Results and Discussion

The compositions of the feed coal, the Flash Pyrolysis char and the acid-leached char are given in Table I. A comparison of the sulfur forms in feed coal and Flash Pyrolysis char shows that the total sulfur was approximately constant and that pyritic sulfur and some organic sulfur were converted to sulfide sulfur. The conversion of the coal pyritic sulfur to sulfide sulfur in the char is beneficial to the AL/HDS process because the sulfide form will readily dissolve in the process acids. The absence of iron sulfides is necessary for the improved hydrodesulfurization rate.

Table I. Coal and Char Composition[a]

	Feed Coal	Flash Pyrolysis Char	Acid-Leached Char
Moisture	3.43	0.66	2.26
Ash (dry)	9.28	24.77	17.97
Volatile matter (dry)	38.15	5.17	12.34
Fixed carbon (dry)	52.57	70.06	69.69
Carbon (dry)	72.35	72.98	76.03
Hydrogen (dry)	4.70	0.83	1.19
Sulfur (dry)	2.96	2.33	1.06
Oxygen and nitrogen (dry)	10.71	0	0
Total sulfur (MAF)	3.26	3.09	1.30
Sulfide sulfur (MAF)	0.13	1.98	0.19
Pyritic sulfur (MAF)	1.44	0.17	0.07
Sulfate sulfur (MAF)	0.09	0	0
Organic sulfur (MAF)	1.60	0.95	1.04
Chloride (dry)	–	0.05	0.33
Iron (dry)	1.32	4.01	0.36
Calcium (dry)	0.31	0.98	0.14
Silicon (dry)	2.36	5.35	5.66
Aluminum (dry)	0.86	2.03	2.03
Potassium (dry)	0.17	0.43	0.43
Sodium (dry)	0.08	0.18	0.18
Magnesium (dry)	–	0.12	0.08
Heating value (BTU/lb, dry)	12,780	11,280	11,540
EPA standard in % sulfur (MAF)		0.90	0.84

[a]Values are wt % except for heating value.

A comparison of the compositions in Table I of the Flash Pyrolysis (FP) char and the acid-leached (AL) char shows that the acid leach removed approximately 90% each of the sulfide sulfur, iron, and calcium along with 30% of the ash. More moles of iron and calcium were removed than moles of sulfur; thus, potential sulfur scavengers in the mineral matter such as oxides were also removed.

Limited hydrodesulfurization experiments were performed with FP char to confirm the improvements in hydrodesulfurization after acid leach. The data for hydrodesulfurization rates of FP char with pure hydrogen at 1600°F and 65 psia are shown in Figure 1 with curves given for total sulfur, sulfide sulfur, and organic sulfur. All three curves gradually decrease at similar rates thus showing that sulfide sulfur and organic sulfur are removed simultaneously. For this FP char, the EPA emission standard of 1.2 lb SO_2/MM Btu is equivalent to 0.90% total sulfur (MAF) which was not achieved after 150 min in pure hydrogen. Hence, for FP chars residence time greater than 150 min in pure hydrogen would be required to meet the EPA standard by hydrodesulfurization alone.

The data for hydrodesulfurization rates of FP char with 0.5% H_2S in hydrogen are shown in Figure 2 with curves given for total sulfur, sulfide sulfur, and organic sulfur. The total sulfur and sulfide sulfur both increased, and the organic sulfur remained approximately constant. In these experiments the H_2S/H_2 ratio was 5×10^{-3}. At 1600°F the accepted experimental equilibrium constant (H_2S/H_2 ratio) is 3×10^{-3} (3) for the reaction of FeS with hydrogen:

$$FeS + H_2 \rightleftharpoons Fe + H_2S$$

Therefore, the 0.5% H_2S in H_2 will force the equilibrium reaction to the left and prevent the removal of sulfide sulfur. The increase in sulfide sulfur confirms the presence of sulfur scavengers in the mineral matter. The reaction mechanisms of organic sulfur removal are too complex to predict the level of H_2S that inhibits hydrodesulfurization. It can be seen in Figure 2 that at a 5×10^{-3} H_2S/H_2 ratio, the lowest used in this study, the removal of organic sulfur is also inhibited in FP char. After 150-min residence time, it appears, in addition, that there is a conversion of sulfide sulfur to organic sulfur as has been observed in the coking of coals (4,5,6).

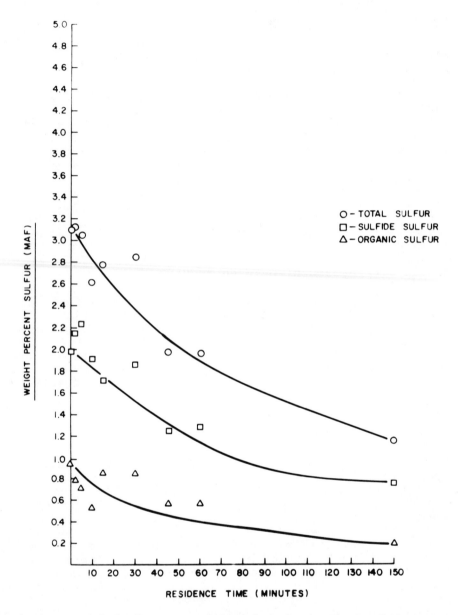

Figure 1. Hydrodesulfurization of flash pyrolysis char at 1600°F and 65 psia in hydrogen

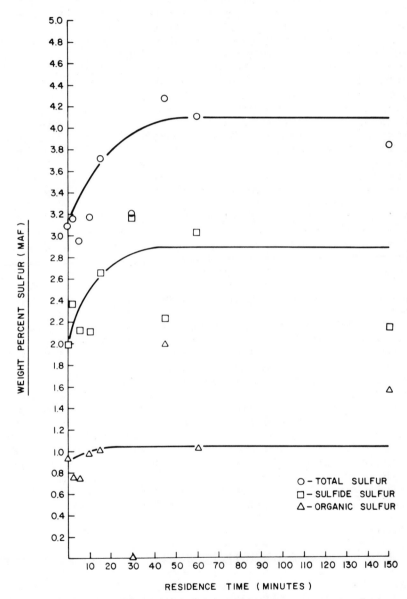

Figure 2. *Hydrodesulfurization of flash pyrolysis char at 1600°F and 65 psia in 0.5% H₂S/BAL H₂*

Since long residence times are usually required for the hydrodesulfurization of coals and chars, the possibility of sulfide sulfur conversion to organic sulfur can impose added restrictions on allowable H_2S in the gas if operative at lower H_2S/H_2 ratios.

The data for hydrodesulfurization rates of AL char at 1600°F and 65 psia with pure hydrogen and five levels of hydrogen sulfide in hydrogen are given in Figure 3. Only the total sulfur curves are shown since they are essentially the same as the organic sulfur curves for AL chars. The sulfur levels reached a steady state after 10 min in all curves except pure hydrogen which required about 30 min. As the hydrogen sulfide concentration in the hydrogen increases, the hydrogen sulfide inhibition of organic sulfur removal is exhibited in the gradually increasing steady-state level of sulfur in the char. From the steady-state levels drawn in these curves, the EPA standard of 0.84% total sulfur (MAF) for AL char can be met with H_2S concentrations as high as 4%. In both the 4.0% and 8.5% H_2S curves, there is the indication that sulfur increased at 150 min residence time as was observed with the FP char and 0.5% H_2S. This again supports the hypothesis that the char's reactivity to the addition of sulfur is increased by long residence times.

The inhibition isotherms are shown in Figure 4 for total sulfur, organic sulfur, and sulfide sulfur in FP char and AL char. The organic sulfur and sulfide sulfur curves for FP char readily show that both forms increased when the acid-leachable minerals and hydrogen sulfide are present. However, the inhibition isotherms for AL char show that organic sulfur is removed and that no sulfide sulfur is formed through back reaction for any of the H_2S/H_2 ratios (0.005 - 0.093) tested. Hence, removing the acid-leachable minerals not only eliminates the scavenging of hydrogen sulfide by the minerals but also makes it possible to achieve effective organic sulfur removal in the presence of high H_2S/H_2 ratios.

Conclusions

Effective hydrodesulfurization of Flash Pyrolysis chars requires essentially H_2S-free hydrogen and greater than 150 min residence time at 1600°F and 65 psia. Effective hydrodesulfurization of acid-leached chars can be achieved with hydrogen mixtures containing up

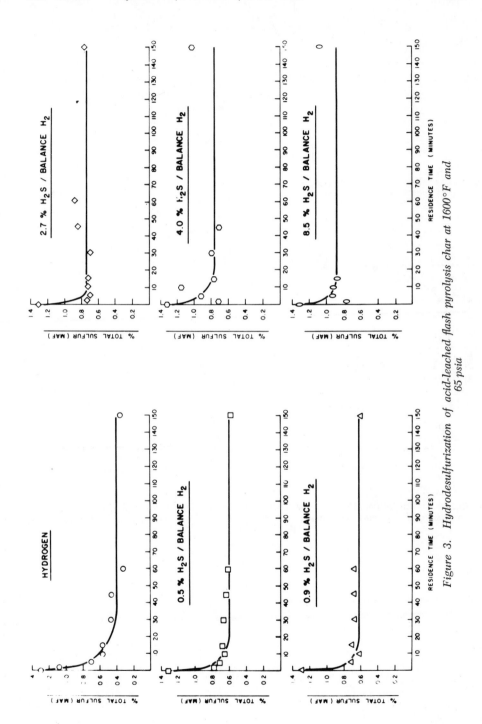

Figure 3. Hydrodesulfurization of acid-leached flash pyrolysis char at 1600°F and 65 psia

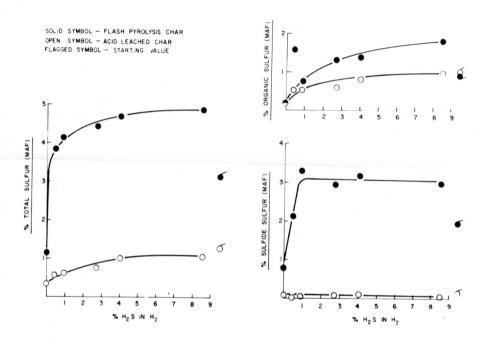

SOLID SYMBOL — FLASH PYROLYSIS CHAR
OPEN SYMBOL — ACID LEACHED CHAR
FLAGGED SYMBOL — STARTING VALUE

Figure 4. Inhibition isotherms for flash pyrolysis char and acid-leached char at 1600°F, 65 psia, and 150 min residence time. (●) flash pyrolysis char, (○) acid–leached char, (●, ○) starting value.

to 4% H_2S in residence times of 10 min or longer at 1600°F and 65 psia.

Literature Cited

1. Sass, A., Chem. Eng. Prog. (1974) 70, 72.
2. Robinson, Leon, Fuel (1976) 55, 193.
3. McKinley, Joseph B., "Catalysis", P.H. Emmett, Ed., Vol. V, p. 453, Reinhold, New York, 1957.
4. Powell, Alfred R., J. Ind. Eng. Chem. (1920) 12, 1069.
5. Eaton, S.E., Hyde, R.W., Old, B.S., Am. Inst. Mine Met. Eng., Iron Steel Div. Metal Tech. (1948) 19, 343.   Tech. Bull. 2453.
6. Cernic-Simic, S., Fuel (1962) 41, 141.

22

Coal Desulfurization during Gaseous Treatment

EDMUND TAO KANG HUANG* and ALLEN H. PULSIFER

Department of Chemical Engineering and Nuclear Engineering and
Energy and Mineral Resources Research Institute,
Iowa State University, Ames, IA 50011

Sulfur removal from coal during treatment with gaseous atmospheres at elevated temperatures is of interest both because such a treatment might serve as a basis for a coal desulfurization process (1,2,3) and because knowledge of the fate of sulfur when coal is heated has application during carbonization or gasification. Early interest in desulfurization under these conditions concentrated mainly on the production of metallurgical coke. The various approaches included pretreatment of the coal before carbonization, carbonization of the coal in various gases to remove sulfur compounds, or treatment of the coke to reduce its sulfur content. Only recently has interest turned to desulfurization of coal or char for use as a power plant fuel.

Sulfur removal from coal during carbonization has been investigated both with and without the addition of reactive gases such as hydrogen and oxygen. During carbonization in the presence of only the gases derived from the coal, hydrogen sulfide is released along with small amounts of other volatile sulfur-containing compounds (4). The iron pyrite decomposes when heated and releases half its sulfur, while one-quarter to one-third of the organic sulfur is converted to hydrogen sulfide (5,6). The total amount of sulfur removed from coal during carbonization is influenced by the coal rank and quantity and composition of the mineral matter (7).

Snow (8) heated coal in various reactive gases and found hydrogen to be the most effective desulfurizing agent of the gases tested. A higher rate of desulfurization and lower char sulfur content were reported when hydrogen was used as compared with normal carbonization, with both quantities being favored by increased hydrogen pressure (9,10). Pyrite reacts with hydrogen in two steps, being first converted to ferrous sulfide and then to pure iron. Yergy et al. (11) found that the first reaction begins at 400°C while the second reaction occurs at a somewhat higher temperature. The conversion of the pyrite to sulfide appears to be fairly rapid while the reaction of hydrogen with ferrous sulfide is slow (12). The conversion of organic sulfur to hydrogen sulfide was reported by Powell (13) not to be significant below 500°C but increases greatly as the temperature is raised from 500° to 1000°C. Hydrodesulfurization of coal is strongly inhibited by the presence of hydrogen sulfide in the gas phase (12,14,15), and Gray et al. (12) found that sulfur removal is rapid and limited by the equilibrium between hydrogen sulfide and the sulfur in the coal. An acid leach of the coal removes iron and calcium sulfides, reducing the back reaction of coal with hydrogen sulfide and significantly improving the effectiveness of hydrodesulfurization (16,17).

Desulfurization of coke and coal with oxygen-containing gases has been the subject of several investigations (18,19,20). Mainly pyritic sulfur is removed in oxidizing atmospheres. Sinha and Walker (18) reported that 90% of the pyritic sulfur was removed in 10 minutes at 450°C with a 5-17% sample weight loss. However, Block et al. (20) found in experiments at similar conditions that less pyritic sulfur was removed with a greater weight loss.

Several investigators have reported that some of the inorganic sulfur in coal is transformed into organic sulfur upon heating (5,7,8,11). Cernic-Simic (7) added radioactive sulfur in the form of iron pyrites to a coal sample and after carbonization found that part of the radioactive sulfur was retained by the coal in the form of organic sulfur. The amount of radioactive sulfur retained as organic sulfur increased as coal rank decreased.

In the work reported here, desulfurization of both a raw and deashed Iowa coal were investigated in three different gases: nitrogen, hydrogen, and a 6% oxygen - 94% nitrogen gas mixture. With each gas, both the temperature and holding time at temperature were varied. With hydrogen and nitrogen, the temperature was varied between 300° and 700°C. The temperature range used with the oxygen-nitrogen mixture was only 350° - 455°C because of the large weight loss at higher temperatures caused by combustion of the sample. Holding time at temperature was varied from 0 to 60 minutes, except in the oxidizing atmosphere where the maximum holding time was 40 minutes.

Experimental

The investigation was carried out with a Rigaku CN8001 H type thermal analyzer which included a thermal gravimetric analyzer, a preprogrammed heating unit and temperature control system, and a data recording unit. The coal sample was confined in a 10-mm diameter platinum pan having a depth of 5 mm. The pan was supported on a balance that could detect a weight change of less than 0.1 mg. The sample holder was contained in a 20-mm diameter quartz tube which was continuously purged with gas.

To carry out a run, about 300 mg of coal (<125 μm) were placed in the sample pan of the thermal analyzer and heated to

the desired temperature at 20°C/minute while maintaining a flow
of gas over the sample of about 1 L/minute. After holding the
sample at temperature for the desired period of time, the gas
flow was switched to nitrogen and the reaction chamber was
cooled with the initial cooling rate being on the order of
250°C/minute. The sample weight was recorded continuously, and
at the end of the run the total and organic sulfur contents of
the treated coal were determined using an oxygen-flask method
(21,22). The organic sulfur content was determined from a
sample that had been leached overnight in 12.5% by volume nitric
acid.

The coal came from the Jude mine, Mahaska County, Iowa, and
is a high-volatile C bituminous coal. The raw coal contained
3.25% inorganic sulfur and 3.04% organic sulfur (moisture-free
basis). The deashed Jude coal was prepared by twice floating
the raw coal in a heavy liquid medium having a specific gravity
of 1.3. It contained 4.74% organic sulfur but only 0.69%
inorganic sulfur. Thus, the results of experiments using the
deashed coal manifested largely the behavior of the organic sulfur compounds in the coal. The composition of the raw and
deashed coal are shown in Table I.

Table I.   Composition of Coal Samples

	Raw Coal	Deashed Coal
Proximate analysis (wt %)		
moisture	1.22	2.21
ash	18.80	5.54
volatile matter	39.68	45.72
fixed carbon	40.30	46.53
Ultimate analysis (wt %, maf)		
carbon	60.45	
hydrogen	4.56	
nitrogen	~ 1	
oxygen	~ 26	
organic sulfur 3.75		5.02
inorganic sulfur 4.01		0.73
total sulfur	7.76	5.75
Heating value (MJ/kg)	25.9	29.4
(Btu/lb)	11,140	12,640

Results and Discussion

The effect of temperature upon sulfur removal was determined
in each of the three gases by heating coal samples to the desired
temperature using a heating rate of 20°C/minute, followed by
immediate cooling of the sample. In addition, the relationship

between desulfurization and holding time was investigated at a number of temperatures. In this case the sample was held at constant temperature for a specified time after being heated to temperature using the same 20°C/minute rate.

Nitrogen. As the temperature of the coal is raised in the presence of nitrogen, volatile matter is lost from the sample. This is accompanied by loss of sulfur. As shown in Figure 1, the amount of all forms of sulfur removed from both the raw and deashed Jude coal increased with increasing temperature, except for the amount of organic sulfur removed from the raw coal. Nearly 80% of the inorganic sulfur was lost by both coals at 700°C. The inorganic sulfur included some sulfates but was primarily pyrite. Since pyrite would be expected to form ferrous sulfide upon heating rather than pure iron (14), the high inorganic sulfur removal was caused presumably by reaction of ferrous sulfide with volatile organic compounds formed from the coal.

The amount of organic sulfur removed from the deashed coal increased as the temperature and hence weight loss increased, with a 36% weight loss being accompanied by a sulfur removal of about 58% at 700°C. With the raw coal, however, the amount of organic sulfur removed first increased and then decreased as the temperature was raised. In fact, the raw coal contained more organic sulfur after being heated to 700°C than it had originally. Since the raw coal, unlike the deashed coal, contained a substantial amount of inorganic sulfur, this behavior seems to confirm the observation by other investigators (5,7,8,11) that inorganic sulfur in coal is transformed into organic sulfur upon heating. The transformation could be caused by direct reaction of pyrite and the carbonaceous material or reaction of the sulfur vapor and hydrogen sulfide produced from pyrite with the organic portion of the coal. The reaction of hydrogen sulfide with carbon-containing materials is rapid according to Kor (10) and Yergy et al. (11).

Also shown in Figure 1 is the total sulfur content of the char. As the temperature was increased, the sulfur content of char produced from the raw coal decreased only slightly indicating that the sulfur loss from the coal was essentially proportional to the weight loss. The sulfur content of the char from the deashed coal, on the other hand, decreased as the temperature was increased.

The sulfur content of the two types of coal did not change significantly when held at constant temperature. Most of the sample weight loss and sulfur removal occurred during the heating step, again indicating that sulfur removal was caused mainly by pyrolysis and release of volatile matter.

Hydrogen. Figure 2 shows the effect of temperature on total and organic sulfur removal from raw and deashed Jude coal when

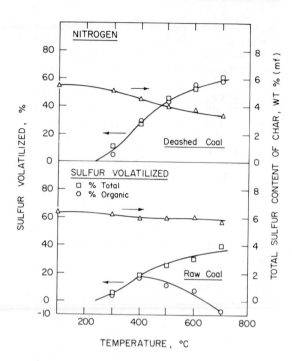

Figure 1. Total sulfur and organic sulfur released in nitrogen and sulfur content of product char

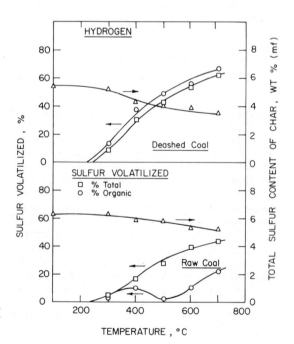

Figure 2. Total sulfur and organic sulfur released in hydrogen and sulfur content for product char

treated with hydrogen. Again, except for the organic sulfur in the raw coal, the amount of all forms of sulfur removed increased as the temperature was increased. Furthermore, the fraction of these forms of sulfur removed in hydrogen at a given temperature generally was within 10% of that lost in nitrogen.

Up to 600°C, the amount of inorganic sulfur removed from the raw coal at any temperature was somewhat higher in hydrogen than in nitrogen. However, at 700°C, about 10% less inorganic sulfur was removed by hydrogen. Since other runs at 700°C and longer holding times were consistent with this result, it does not seem likely that this observation resulted from experimental error. Yergy et al. (11) reported that the onset of the hydrogen sulfide-iron reaction occurs at 530°C, and it is possible that this reaction became significant and led to some conversion of pure iron back to the sulfide form between 600° and 700°C.

With the deashed coal, the amount of organic sulfur lost in hydrogen was generally 5-10% above that lost in nitrogen, while the weight loss in the two gases differed by 2-6% with the loss in hydrogen being higher. The effect of temperature on removal of organic sulfur from the raw coal, however, was quite different in hydrogen than nitrogen. The amount of organic sulfur removed from the raw coal in hydrogen first increased and then decreased as in nitrogen. However, the amount increased again at temperatures above 500°C. Hydrodesulfurization, therefore, only seems to become effective above 500°C. The decrease in organic sulfur removed between 400° and 500°C suggests that inorganic sulfur also was transformed into organic sulfur in the presence of hydrogen.

The sulfur content of the product char decreased somewhat as the coal was treated at higher temperatures (Figure 2). The decrease for both the raw and deashed coal in hydrogen was slightly larger than that observed with nitrogen.

Figure 3 shows the effect of holding time upon the amount of sulfur removed in hydrogen. For both temperatures shown, 400° and 700°C, the coal sample was heated to the specified temperatures at 20°C/minute and then was held there for the time shown. At 400°C, the amount of sulfur removed increased slightly at first, then remained fairly constant as the sample was held at temperature. During this time, the total sulfur content of the chars produced from the two coals remained nearly constant. At 700°C, the amount of sulfur removed from the coal increased continuously as the holding time was increased. The total sulfur content of the char also decreased continuously, with the sulfur content of the chars produced from the raw and deashed coals being reduced from 5.26 to 3.79% and 3.54 to 2.43%, respectively, when held in hydrogen for 60 minutes. Desulfurization in hydrogen, then, was more effective at the higher temperature.

The major differences between desulfurization in nitrogen and hydrogen seemed to occur in the data taken at 600° and 700°C. An examination of the organic sulfur removed versus weight loss

curves for deashed coal clearly demonstrates this (Figure 4).
Since removal of organic sulfur in nitrogen primarily results
from the evolution of volatile compounds containing sulfur, the
organic sulfur removed and weight loss should be related by a
single curve irrespective of temperature and treatment time.
As can be seen in Figure 4, this is indeed the case as all of
the data points collected in nitrogen at the various temperatures and holding times fell on a single curve.

The organic-sulfur-removed-weight-loss relationship also
was a single, smooth curve in hydrogen for all temperatures and
holding times (Figure 4). Up to a weight loss of about 40%
(point A, Figure 4), the curve in hydrogen was nearly the same
as in nitrogen. However, at higher weight losses, the organic
sulfur removed in hydrogen increased rapidly with only a slight
increase in weight loss. All the experimental points at weight
losses greater than that shown by point A were taken at 600° and
700°C. The reaction of hydrogen with sulfur in coal appears,
therefore, to be significant only at 600°C or higher. At lower
temperatures, desulfurization in both hydrogen and nitrogen seems
to occur mainly by removal of volatile, sulfur-containing
compounds.

Figure 4 shows only data collected with deashed Jude coal.
The organic sulfur removed from the raw coal did not correlate
with weight loss because of the transformation of inorganic
into organic sulfur which took place during the heating process.

Oxygen-Nitrogen Mixture. The final gas tested was nitrogen
containing 6% oxygen. The maximum treatment temperature in this
gas mixture was limited to 455°C because of oxidation of the
sample and the resulting large weight loss as the temperature
was increased. The weight loss in the oxidizing atmosphere at
455°C was about the same as in the other gases at 700°C. The
amount of all forms of sulfur removed from the deashed coal and
the total and inorganic sulfur removed from the raw coal
increased with increasing temperature as before (Figure 5).
The amount of organic sulfur removed from the raw coal
increased, then decreased, and finally increased again
between 400° and 455°C. This latter behavior seems to
suggest that transformation of the inorganic into organic
sulfur again was taking place. The sulfur content
of the chars produced from both coal samples decreased fairly
significantly as the temperature was increased, with most of the
decrease occurring as the coal was heated from room temperature
to 350°C.

The percentage of sulfur removed in the oxygen-containing
gas at a given temperature was always higher than in hydrogen
or nitrogen. This is partly caused by the higher weight loss
experienced with oxygen, and partly because more pyritic sulfur
was removed in the oxidizing atmosphere. Pyritic sulfur removal
is known to be favored in oxygen (14,15,18).

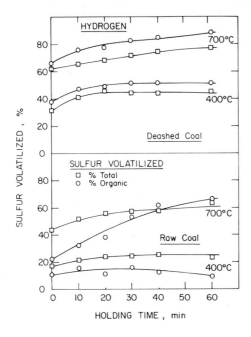

Figure 3. Effect of holding time on total sulfur and organic sulfur removal in hydrogen

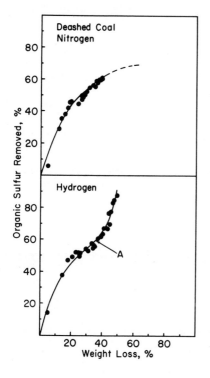

Figure 4. Organic sulfur removal as a function of sample weight loss for deashed Jude coal

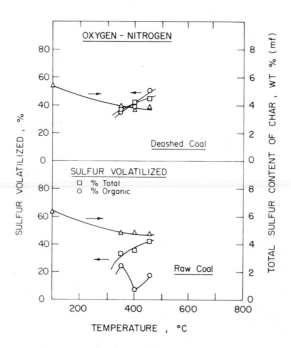

Figure 5. Total sulfur and organic sulfur released in nitrogen–6% oxygen mixture and sulfur content of product char

tment condition, the material balance equations allowed
of f, G_1, and G_2 to be made for all temperatures and
imes used. Since there are four equations, one of the
s may be calculated twice. Two independent estimates
e made for each experimental condition, and in all
se were essentially the same.
es of f, G_1, and G_2 for each of the three gases are
Table III. The fixation reaction, characterized by f,
t 300°C in nitrogen and increased rapidly between 400°
with f reaching 0.7 at 700°C. G_1 was generally small
peratures, being between 0 and 0.1. G_2 increased rapidly
)0° and 500°C and then started to level off between
'00°C. In nitrogen, the effect on G_1, G_2, and f of
me when the sample was kept at constant temperature

raction of inorganic sulfur transformed into organic
was larger in hydrogen than in nitrogen at low
es. However, f remained fairly constant at tempera-
e 400°C. G_1 was again small while G_2 increased with
temperature as it had in nitrogen. f, G_1, and G_2
elatively constant when the samples were held at
es less than 700°C. At 700°C, G_1 and G_2 both
slightly with holding time, but f decreased from 0.5
.2 when the holding time was varied from 0 to 60
Apparently the hydrogen sulfide formed from the organic
ted with the iron to produce ferrous sulfide. As
noted, such a reaction had been found to occur at
atures (11,16).
action of inorganic sulfur transformed into organic
in the oxidizing atmosphere was comparable with that
at the same temperature. It increased rapidly from
350°C to 0.36 at 455°C. The fraction of inorganic
ased into the gas phase (G_1) was much larger than in
hydrogen since the reaction of pyrite with oxygen
id than the reaction of pyrite with hydrogen and was
anged with temperature between 350° and 455°C. G_2
rger in the oxygen-nitrogen mixture and increased
temperature. About 55% of the organic sulfur was
455°C. G_1 and G_2 both increased with holding time
the oxidizing atmosphere while f decreased as the
was increased. The decrease in f is presumably
idation and release of sulfur that had previously
rmed from inorganic to organic sulfur which would
crease in f and an increase in G_1.
ction of inorganic sulfur transformed into an organic
generally large under all conditions tested. Cernic-
orted that the fraction transformed varies with coal rank
ansformation of sulfur taking place with lower rank coal.
he value of f for Jude coal at a given condition pro-
end to be above the average for coals taken as a

Data were taken on the effect of holding time in the oxygen-
nitrogen mixture at 350°, 400°, and 455°C. At each of the three
temperatures, the amount of sulfur removed increased significantly as the holding time was increased from 0 to 40 minutes.
This was accompanied by an increased weight loss. There was,
however, some preferential removal of the sulfur as the total
sulfur content of the product char decreased with increased
holding time in each case. For example, the total sulfur
content of the char produced from the raw coal was reduced by
about 0.5 wt% when the sample was held at temperature for 40
minutes.

Transformation of Sulfur. Since the data indicated that
some of the inorganic sulfur was trapped by the organic portion
of the coal during heating, an attempt was made to more quantitatively estimate the fate of the inorganic and organic sulfur
in the coal during the heating process. The following reaction
scheme first was postulated where G_1 is the fraction of inorganic
sulfur released, f is the fraction of inorganic sulfur
transformed into organic sulfur, and G_2 is the fraction of original organic sulfur released as gas.

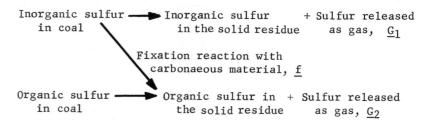

Assuming that this reaction scheme applies to both the raw
and deashed coal with the same values of f, G_1, and G_2 at any
given condition, the parameters can be estimated from sulfur
material balances, as shown in Table II.

Table II. Sulfur Material Balances

Raw Coal

$$3.25 - S_i = 3.25\ f + 3.25\ G_1 \quad \text{(inorganic sulfur)}$$
$$3.04 - S_o = 3.04\ G_2 - 3.25\ f \quad \text{(organic sulfur)}$$

Deashed Coal

$$0.69 - S_i' = 0.69\ f + 0.69\ G_1 \quad \text{(inorganic sulfur)}$$
$$4.74 - S_o' = 4.74\ G_2 - 0.69\ f \quad \text{(organic sulfur)}$$

where S_i, S_i' = inorganic sulfur in char, gm/100gm original coal,

S_o, S_o' = organic sulfur in char, gm/100gm original coal.

Since the inorganic and organic sulfur contents of the chars
produced from both the raw and deashed coals were determined at

Table III. Calculated Parameters for Sulfur Reactions

Temp. (^{o}C)	Holding Time (min)	f	G_1	G_2
Hydrogen atmosphere				
300	0	0.10	-0.05	0.15
400	0	0.29	-0.06	0.42
400	10	0.34	-0.06	0.52
400	20	0.41	-0.05	0.55
400	30	0.38	-0.07	0.57
400	40	0.42	-0.05	0.58
400	60	0.45	-0.08	0.58
500	0	0.51	0.02	0.58
500	10	0.51	0.02	0.61
500	20	0.43	0.22	0.58
500	30	0.39	0.12	0.62
500	40	0.34	0.20	0.57
500	60	0.34	0.23	0.59
600	0	0.49	0.12	0.62
600	10	0.48	0.13	0.68
600	20	0.52	0.19	0.71
600	30	0.57	0.15	0.75
600	40	0.49	0.14	0.73
600	60	0.49	0.09	0.76
700	0	0.49	0.09	0.74
700	10	0.47	0.10	0.83
700	20	0.43	0.17	0.84
700	30	0.32	0.15	0.87
700	40	0.25	0.19	0.88
700	60	0.23	0.29	0.90

Table III. Continued.....

Temp. (^{o}C)	Holding Time (min)	f
Nitrogen atmosphere		
300	0	0.01
400	0	0.12
400	5	0.16
400	10	0.15
400	15	0.19
400	20	0.23
500	0	0.36
600	0	0.50
700	0	0.71
700	5	0.75
700	10	0.76
700	20	0.70
700	40	0.72
Oxygen-nitrogen gas mixture		
350	0	0.12
350	20	0.17
350	40	0.12
400	0	0.36
400	10	0.36
400	20	0.27
400	40	0.0
455	0	0.3
455	20	0.3
455	40	0.3

each tr
estimat
holding
paramet
of G_2 we
cases th
Va
shown in
started
and 500°
at all t
between
600° and
holding
was smal
The
sulfur (
temperat
tures ab
increasi
remained
temperatu
increased
to about
minutes.
sulfur re
previousl
high temp
The
sulfur (f
in hydrog
about 0.1
sulfur re
nitrogen
is more ra
almost un
also was l
rapidly wi
released a
at 400°C i
holding ti
caused by
been trans
lead to a
The fr
form (f) wa
Simic (7) re
with more t
Therefore,
bably would
whole.

Conclusions

Temperature is an important factor influencing desulfurization in all three gases tested. The holding time at temperature is less important except in the case of oxidizing atmospheres or hydrogen above 600°C where the sulfur content of the sample decreases with increased holding time.

The mechanism and extent of desulfurization also depends upon the gaseous atmosphere. Sulfur removal in nitrogen and in hydrogen below 600°C is caused mainly by pyrolysis and release of volatile matter. Hydrogen reacts directly with sulfur in the organic portion of the coal only at 600°C or above. Pyritic sulfur removal is greater in oxidizing atmospheres.

In the experiments some inorganic sulfur was transformed into organic sulfur during the gaseous treatment. The fractions of inorganic and organic sulfur released as gases and the fraction of inorganic sulfur transformed into organic sulfur were estimated for each of the gas atmospheres and conditions of temperature and time used.

Acknowledgement

Work sponsored by the Iowa Coal Project and conducted in the Energy and Mineral Resources Research Institute at Iowa State University. Dr. E. T. K. Huang is currently at Allis-Chalmers, Milwaukee, Wisconsin.

Literature Cited

1. Haldipur, G., Wheelock, T. D., this volume.
2. Boodman, N. S., Johnson, T. F., Krupinski, K. C., this volume.
3. Fleming, D. K., Smith, R. D., Aquino, M. R. Y., this volume.
4. Thiessen, G. "Pyrolytic Reactions of Coal," Chem. Coal Util. Suppl. Vol. (1963) 340-394.
5. Powell, A. R., Ind. Eng. Chem. (1920) 12, 1069-1077.
6. Powell, A. R., J. Am. Chem. Soc. (1923) 45, 1-15.
7. Cernic-Simic, S., Fuel (1962) 41, 141-151.
8. Snow, R. D., Ind. Eng. Chem. (1932) 24, 903-909.
9. Zielke, C. W., Curran, G. P., Gorin, E., Goring, G. E., Ind. Eng. Chem. (1954) 46, 53-56.
10. Kor, G. J. W., this volume.
11. Yergy, A. L., Lampe, F. W., Vestal, M. L., Day, A. G., Fergusson, G. L., Johnston, W. H., Snyderman, J. S., Essenhigh, R. H., Hudson, J. E., IEC Proc. Res. Dev. (1974) 13, 233-240.
12. Gray, C. A., Sacks, M. E., Eddinger, R. T., IEC Prod. Res. Dev. (1970) 9, 357-361.
13. Powell, A. R., Ind. Eng. Chem. (1920) 12, 1077-1081.

14. Maa, P. S., Lewis, C. R., Hamrin, C. E., _Fuel_ (1975) 54, 62-69.
15. Batchelor, J. D., Gorin, E., Zielke, C. W., _Ind. Eng. Chem._ (1960) 52, 161-168.
16. Robinson, L., _Fuel_ (1976) 55, 110-118.
17. Tipton, A. C., this volume.
18. Sinha, R. K., Walker, P. L., _Fuel_ (1972) 51, 125-129.
19. _Ibid._, 329-331.
20. Block, S. S., Sharp, J. B., Darlage, L. J., _Fuel_ (1975) 54, 113-120.
21. Ahmed, S. M., Whalley, F. J. P., _Fuel_ (1972) 51, 190-193.
22. _Ibid._, (1974) 53, 61-62.

Desulfurization of Coal in a Fluidized-Bed Reactor

G. B. HALDIPUR and T. D. WHEELOCK

Department of Chemical Engineering and Nuclear Engineering,
Energy and Mineral Resources Research Institute,
Iowa State University, Ames, IA 50011

The pioneering investigation of Jacobs and Mirkus (1) showed that substantial amounts of sulfur could be removed from Illinois No. 6 coal by treatment with mixtures of air, nitrogen, and steam in a fluidized bed reactor at moderately elevated temperatures. Thus by treating coal, which had been ground in a hammer mill (100% through 8-mesh screen), with a gas mixture containing 2.7% oxygen, 35% steam, and 62.3% nitrogen at 510°C for 30 min, the sulfate and pyritic sulfur content of the solids was reduced about 80% and the organic sulfur content 10%. However, at the same time the content of combustible volatile matter was reduced about 65%. Desulfurization improved with increasing residence time and decreasing particle size, but it was affected only slightly by oxygen concentrations in the range of 2-10% or steam concentrations in the range of 0-85%. The sulfur content of the char declined as the treatment temperature was increased to 430°C, but higher temperatures were not beneficial because desulfurization was accompanied by increased gasification and reduced yield of char.

Even more encouraging results were reported by Sinha and Walker (2) who were able to remove a large percentage of the pyritic sulfur from most of the samples in a series of powdered bituminous coals by treating them in a combustion boat with air at 450°C for 10 min. Moreover, the low and medium volatile bituminous coals in the series only experienced about a 5% weight loss and the high volatile bituminous coals a 10-17% weight loss. However, the results of a similar series of experiments by Block et al. (3) were less promising because less pyritic sulfur was removed, and a greater weight loss was incurred.

Although the selective oxidation of pyritic sulfur appeared to play an important role in the foregoing demonstrations of desulfurization, it may not have been an exclusive role because sulfur could also have been removed through pyrolysis and reaction with hydrogen which was released by the pyrolytic decomposition of coal. Numerous studies have shown that part of the sulfur in coal is removed during carbonization and that the addition of hydrogen or carbonization in a stream of hydrogen assists the removal of sulfur, particularly at higher temperatures (3,4,5,6). Under such conditions sulfur is removed principally as hydrogen sulfide. An investigation of coal hydrodesulfurization by a nonisothermal kinetic method revealed several peaks in the evolution rate of hydrogen sulfide. Yergey et al. (7) attributed the first peak, which occurred in the range of 390°-470°C for different coals, to the reaction of hydrogen with two forms of organic sulfur, the second peak at 520°C to the reaction of hydrogen with pyrite, the third peak at 620°C to the reaction of hydrogen with ferrous sulfide (produced by the hydrodesulfurization of pyrite), and the fourth peak to the reaction of hydrogen with a third form of organic sulfur. Unfortunately the hydrodesulfurization of coal is inhibited by hydrogen sulfide in the gas phase which severely limits the concentration build-up of hydrogen sulfide (5,8,9,10).

The work reported here was undertaken to determine the feasibility of desulfurizing a high-sulfur bituminous coal from an Iowa mine by treatment at moderately elevated temperatures in a fluidized bed reactor with either oxidizing, neutral, or reducing gases. Nearly isothermal experiments were carried out with a small fluidized bed reactor to determine the extent of desulfurization and coal weight loss for different conditions of temperature and gas composition. Also the treatments were applied to both run-of-mine coal and beneficiated coal. In addition the off-gas composition was measured during some experiments to determine the distribution of various sulfur and other compounds and to estimate the heating value of the gas. Finally consideration was given to the possibility of desulfurizing the off-gas and using it as a clean fuel to burn along with partially desulfurized coal char in the same plant in order to meet air pollution control regulations.

Experimental Investigation

Apparatus. Figure 1 is a flow diagram of the apparatus used for this investigation. Feed gases were conducted through rotameters, combined, and heated to the reaction temperature by an electric preheater. The hot gas then was passed through a fluidized bed reactor containing the coal being treated and then was conducted to a glass cyclone separator which removed any fine particles of coal elutriated from the bed. The gas was cooled next to condense tar and moisture, filtered with glass wool, and bubbled through an alkaline solution of hydrogen peroxide to remove sulfurous gases. Samples of gas were analyzed periodically with a magnetic type mass spectrometer (Model MS10, Associated Electrical Industries Ltd.).

The reactor was constructed from 2-in. i.d. stainless steel pipe and had an overall length of 18 in. It was fitted with a porous sintered stainless steel gas distributor having an

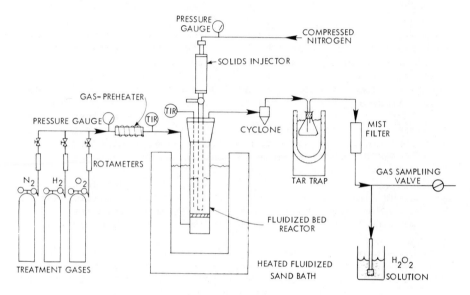

Figure 1. Experimental apparatus flowsheet

effective pore size of 20μm. It was also equipped with a thermo-
well and a device for injecting coal at a point just above the
gas distributor. The reactor was placed in an electrically heated, fluidized sand bath for temperature control.

   Procedure. The reactor was charged with a weighed amount of
-40+50 mesh silica sand. The reactor was then brought up to
operating temperature while air was used as the fluidizing medium.
As the system approached the desired temperature, air was replaced
with the appropriate treatment gas. When the temperature of the
system appeared to have reached a steady state, powdered coal
(-20+40 mesh) was injected into the fluidized bed of sand. This
was done by first filling the injector tube with a weighed amount
of coal. The tube was subsequently pressurized with nitrogen,
and then the quick opening ball valve between the tube and the
reactor was opened allowing the coal to be discharged into the
reactor. This marked the beginning of a run. During a run, the
gas flow through the reactor and the temperature of the fluidized
sand bath surrounding the reactor were kept constant. During
some runs, samples of the off-gas were collected in glass bulbs
at discrete time intervals and later were analyzed with the mass
spectrometer. After a run was completed, the reactor was un-
coupled and doused with water to cool it to room temperature.
The contents of the reactor were weighed and screened to separate
the sand and coal char. The proximate analysis, heating value,
and sulfur distribution of the char were subsequently determined
by the ASTM method. This method of analysis did not distinguish
between sulfur present as ferrous sulfide (FeS) and organic
sulfur.

   Materials. Two run-of-mine (ROM) samples of high volatile
C bituminous coal from the Jude Coal Co. strip mine in Mahaska
County, Iowa, were treated. The samples were crushed and screened to provide material in the -20+40 mesh size range. After
sieving, each sample was split into two fractions. One fraction
was used as is, while the other fraction was beneficiated by a
float/sink technique using a liquid medium (a mixture of hexane
and tetrachloroethylene) having a specific gravity of 1.30.
Since this method of beneficiation greatly reduced the ash content as well as the pyritic sulfur content of the coal, the
beneficiated fraction is referred to as deashed coal. The composition and heating value of the two run-of-mine samples and
corresponding deashed fractions are shown in Table I.

Results and Discussion

   First Series of Runs. The first series of runs was carried
out to determine the effects of four different treatment gas compositions and three different temperature levels (240°, 325°, and
400°C) on the desulfurization of both run-of-mine coal and deashed
coal. The treatment gases included (1) 100% N_2, (2) 85% H_2, 15%
N_2, (3) 4% O_2, 96% N_2, and (4) 10% O_2, 90% N_2. Coal identified

Table I. Composition of Jude Mine Coal on an As-Received Basis

Type of Analysis	Sample I		Sample II	
	ROM	Deashed	ROM	Deashed
Proximate (wt %)				
moisture	6.35	2.24	5.37	4.04
volatile matter	41.14	46.03	40.61	45.60
fixed carbon	38.68	48.84	39.41	47.50
ash	13.83	2.90	14.61	2.86
Sulfur (wt %)				
sulfate	0.49	0.39	0.76	0.38
pyritic	2.40	0.60	2.87	0.60
organic	3.54	3.97	4.43	5.37
total	6.43	4.96	8.06	6.35
Heating value (Btu/lb)	10,980	13,430	10,860	12,990
Specific sulfur content (lb. $S/10^6$ Btu)	5.86	3.69	7.42	4.89

as Sample I in Table I was used for this series. For each run, 50 g of coal was injected into 400 g of silica sand fluidized with the appropriate treatment gas at a superficial velocity of 30–40 cm/sec. As soon as the coal was added, the temperature of the fluidized bed in the reactor dropped 15°–50°C. However, the temperature of the bed recovered to its initial temperature in 5–10 min and then remained constant for the remainder of a run except for runs made at the highest temperature and oxygen levels. For these runs, the temperature of the bed continued to rise throughout a run so the final temperature was 60°–70°C higher than the initial temperature. This increase in temperature seemed due to partial combustion of the coal or its decomposition products. Each run lasted 30 min. For this series of runs only the char product was recovered and analyzed; the off-gas was not sampled.

The results of runs made with Sample I, run-of-mine coal are presented in Table II. Since duplicate runs were made at the lowest and highest temperatures, each listed value represents an average for two runs at these temperatures. On the other hand, each listed value for the intermediate temperature level represents the result of a single run. During each run the coal experienced some loss in weight because of the escape of volatile matter. This loss increased directly with temperature but was

Table II. Results of the First Series of Runs with Run-of-Mine
           Coal

| Trt. Gas | Temp. (°C) | Wt. Loss (%) | Sulfur Removed (%) | | | lb. S^a / 10^6 Btu |
			Pyritic	Organic	Total	
100% N_2	235	11.6	9.2	10.7	7.4	6.3
	320	14.0	7.8	3.2	2.5	6.6
	400	31.6	7.4	49.1	29.1	6.4
85% H_2	235	11.8	7.1	7.0	6.6	6.1
	325	15.0	7.7	12.3	10.1	6.0
	400	33.6	29.2	35.4	39.7	5.5
4% O_2	235	11.7	8.2	12.2	6.7	6.4
	320	16.0	12.9	25.3	19.1	5.8
	410	30.7	41.2	46.4	45.7	4.9
10% O_2	240	10.0	7.9	18.4	11.3	6.2
	330	18.5	8.6	22.7	11.3	7.3
	440	63.0	73.3	79.8	77.9	5.6

^{a}Specific sulfur content of char product.

not much different for different treatment gases except for the
case when a gas containing 10% oxygen was used at the highest
temperature, and over 60% of the coal was consumed. With this
one exception the weight loss seemed caused primarily by pyrolysis
rather than by reactions involving any of the treatment gases,
although the volatile decomposition products were obviously not
the same for different treatment gases. Thus some black tar was
condensed from the off-gas when either nitrogen or hydrogen was
used, and only a small amount of light oil and water was condensed
when either of the oxygen bearing gases was used.

The percentage of either pyritic, organic, or total sulfur
removed from the coal was determined as follows:

$$\text{Desulfurization (\%)} = \frac{\text{S wt. in feed} - \text{S wt. in product}}{\text{S wt. in feed}} \times 100 \quad (1)$$

Only a small percentage of the pyritic sulfur was removed at any
of the temperature levels when pure nitrogen was used as the
treatment gas (Table II). However, when either hydrogen- or
oxygen-bearing gases were used, a significant percentage of the
pyritic sulfur was removed at the highest temperature with more
sulfur being extracted by oxygen than by hydrogen. The percentage of organic sulfur removed was strongly affected by temperature, but it was affected very little by the treatment gas composition even though it may have appeared that more organic sulfur was removed at 400°C by either nitrogen alone or oxygen-
nitrogen mixtures than by hydrogen. A qualitative chemical

analysis showed that some of the "organic" sulfur in char produced during the runs with hydrogen was actually an inorganic
sulfide. A similar analysis of the char produced during the runs
with oxygen in the feed gas did not reveal any sulfide. Furthermore, so little pyritic sulfur was removed during the runs with
pure nitrogen that not much sulfide could have been produced.
Therefore only the results from the hydrogen runs are suspect,
and the organic sulfur removed at 400°C was probably greater
than indicated because of this problem with the chemical
analysis. Considering that the removal of organic sulfur depends
strongly on temperature and very little on treatment gas composition, it appears that such removal is caused mainly by
pyrolysis and release of volatile matter.

The cumulative distribution of various forms of sulfur remaining in either run-of-mine or deashed coal after treatment
with oxygen bearing gases is shown in Figure 2. The vertical
distance separating any given pair of curves represents the percentage of the indicated species of sulfur found in the product
based on the total sulfur in the feed, and it was determined by
using the relation:

$$S \text{ species } (\%) = \frac{\text{wt. of species in product}}{\text{total wt. of S in feed}} \times 100 \qquad (2)$$

The distribution at the left-hand side of each diagram corresponds
to the sulfur distribution of the feed material. A comparison of
the sulfur distribution at different temperatures with the
initial distribution shows that for every treatment gas, the total
amount of sulfur remaining in the solids decreased as the temperature was raised with the greatest change generally taking place
above 325°C. In the case of either run-of-mine or deashed coal
treated with oxygen, both organic and inorganic sulfur were removed, but at higher temperatures more inorganic sulfur appeared
to be removed than organic relative to the amount of each species
present initially.

The sulfur distribution diagrams also indicate the inter-
conversion between forms of sulfur. Thus it appears that the
sulfate form of sulfur gained slightly at the expense of other
forms of sulfur when run-of-mine coal was treated with an oxygen-
bearing gas at 235°C. However, it does not appear that any of
the treatments produced a wholesale transformation of one form
of sulfur into another. There certainly was little if any
evidence such as Cernic-Simic (4) had found indicating the
transformation of organic sulfur into inorganic sulfur.

As a result of volatile matter loss and/or coal oxidation
which accompanied desulfurization, the specific sulfur content
(pounds of sulfur per million Btu) of the coal was not reduced
materially by any of the treatments. In fact for most of the
treatments, the specific sulfur content of the treated run-of-
mine coal (Table II) was actually slightly larger than that of
the feed (5.86 lb. S/10⁶ Btu). For run-of-mine coal the lowest

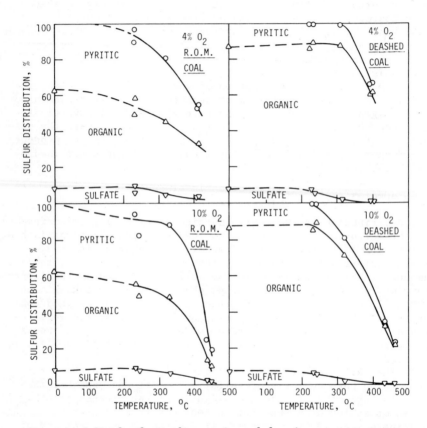

Figure 2.   Sulfur distribution diagrams for coal char after oxygen treatments

specific sulfur content (4.9 1b. S/10^6 Btu) was obtained when it was treated at 410°C with gas containing 4% oxygen. For deashed coal the specific sulfur content of the product was slightly less than that of the feed (3.69 1b. S/10^6 Btu) following most of the treatments, and at the highest temperature level the specific sulfur content of the product was almost the same regardless of treatment gas.

Second Series of Runs. The second series of runs was conducted to measure the yield and composition of the gaseous reaction product as well as the extent of sulfur removal from both run-of-mine coal and deashed coal. The treatment gases included pure nitrogen and two component mixtures of nitrogen and either hydrogen or oxygen. Coal identified as Sample II in Table I was used for this series. For each run 200 g of coal was injected into 250 g of silica sand fluidized with the appropriate treatment gas at a superficial velocity of 25-50 cm/sec. As soon as the coal was added, the temperature of the fluidized bed in the reactor dropped 115°-170°C. The temperature of the bed usually recovered in 10-15 min to somewhere near its initial value and then remained constant for the duration of a run, except for the runs made with an oxygen-bearing gas where the temperature continued to rise slowly. The runs lasted either 60 or 90 min. For this series of runs the overall yield of liquid condensate was determined, and samples of reactor off-gas were drawn periodically and analyzed with the mass spectrometer. The heating value of the fuel gas portion of the off-gas was estimated by summing the heats of combustion of the individual components. However, for runs using hydrogen as the treatment gas, the contribution of hydrogen to the heating value was excluded.

The results of selected runs in this series of experiments are presented in Table III. Since these runs were made at relatively high temperatures (370°-400°C) and were of long duration, appreciable amounts of volatile matter and sulfur were removed from the coal. When either nitrogen or hydrogen was used as the treatment gas, the off-gas contained small but significant amounts of ethane and propane. A significant amount of hydrogen was also found in the off-gas when pure nitrogen was fed to the reactor. For the runs made with an oxygen-bearing treatment gas, the off-gas contained several percent each of oxygen, carbon dioxide, and carbon monoxide; slightly less hydrogen; a small amount of methane; and trace amounts of ethane and propane. In addition the off-gas contained small amounts of sulfur dioxide and hydrogen sulfide with the former usually exceeding the latter. Traces of carbonyl sulfide were also observed in oxidizing runs. An overall material balance made for each of the selected runs accounted for 97.5-99.9% of all the materials entering and leaving the system.

During each run, the total quantity of sulfur in the off-gas was also determined by absorption and oxidation of the various sulfurous gases in an alkaline solution of hydrogen peroxide, and

Table III.   Results of Selected Runs in Second Series

Run No.	Coal Type	Trt. Gas	Temp. (°C)	Gas vel. (cm/sec)	Time (min)	Wt. loss (%)	Total S removed (%)[a]
MSN-1	ROM	100% N_2	375	44	60	32.8	39.1
MSN-4	deashed	100% N_2	395	26	60	23.5	41.8
MSH-1	ROM	87% H_2	395	48	60	29.8	44.1
MSH-3	deashed	84% H_2	400	32	90	22.4	32.3
MSO-7	ROM	10% O_2	375	34	90	37.4	48.7
MSO-8	deashed	10% O_2	370	26	90	30.1	41.7

Run No.	Liq. yield (lb./lb. coal)[b]	Net fuel gas[c]		Specific sulfur content (lb. S/10^6 Btu)		
		Yield (SCF/lb. coal)	Heat. value (Btu/SCF)	Feed	Char	Char & Gas
MSN-1	0.14	2.04	522	7.4	6.8	6.1
MSN-4	0.14	1.49	524	4.9	3.5	3.3
MSH-1	0.17	0.97	780	7.4	5.9	5.5
MSH-3	0.15	0.96	912	4.9	4.1	3.8
MSO-7	0.10	13.03	432	7.4	6.9	4.5
MSO-8	0.12	7.29	379	4.9	4.2	3.6

[a] Determined by Equation 1.

[b] Condensed tar and water.

[c] Volume of H_2, CO, CH_4, C_2H_6, and C_3H_8 in off-gas at standard conditions (0°C and 1 atm.), except Runs MSH-1 and MSH-3 where H_2 is excluded.

this quantity agreed reasonably well with the gas analysis made with the mass spectrometer. However, the quantity of sulfur appearing as noncondensible gaseous species was only 40-80% of the sulfur lost by the coal. Hence, the condensed tar and water must have contained an appreciable part of the sulfur extracted from the coal.

For the runs made with hydrogen or nitrogen, the heating value of the coal-derived combustible components in the off-gas was equivalent to 6-11% of the heating value of the char, and for the runs made with an oxygen bearing gas, the heating value of these components was equivalent to 14-36% of the heating value of the corresponding char. Consequently the combined heating value of the char and coal-derived gas was significantly larger than that of the char alone.

The specific sulfur content of both the product char and the char and fuel gas combined was estimated (Table III). For this purpose it was assumed that the off-gas could be completely desulfurized. The specific sulfur content of the char produced during each of the selected runs was significantly less than that of the feed. Furthermore by lumping the char and desulfurized off-gas together, the specific sulfur content of the combined products would be even lower. Thus for the conditions of Run MSO-7 the specific sulfur content of the char was 7% less than that of the run-of-mine coal, and the specific sulfur content of the char and desulfurized gas together would be 39% less. The results of Run MSN-4 indicate the possibility for a 56% overall reduction in the specific sulfur content of the fuel by first beneficiating it and then applying a mild pyrolysis treatment as in this run.

Formation Rates of H_2S and SO_2. The rates of formation of hydrogen sulfide and sulfur dioxide during the second series of runs were estimated by analyzing the time-varying composition of the reactor off-gas as determined by the mass spectrometer. The formation rate of hydrogen sulfide as a function of the conversion of coal sulfur into hydrogen sulfide and sulfur dioxide is shown for several runs made with nitrogen in Figure 3 and for several runs made with hydrogen in Figure 4. Hydrogen sulfide was the principal noncondensible sulfur compound in the off-gas during these runs. For both treatment gases, the formation rate of hydrogen sulfide first increased, subsequently peaked, and then decreased monotonically with increasing conversion. The initial increase in the rate was probably caused by the rise in temperature of the coal after it was first placed in the reactor, and the later decrease in the rate to the diminishing concentration of sulfur in the coal. After it peaked, the rate for deashed coal appeared to be essentially a linear function of the conversion which corresponds to a first-order process. Since the sulfur in deashed coal was present mainly as organic sulfur, this result indicates that the conversion of organic sulfur to hydrogen sulfide is an apparent first-order reaction with respect to the

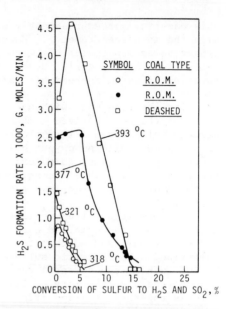

Figure 3.   Rate of H₂S formation during pyrolysis in nitrogen

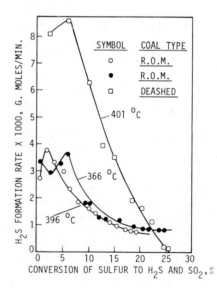

Figure 4.   Rate of H₂S formation during treatment with gas containing 85% hydrogen

sulfur species in coal, which is in agreement with Yergey et al.
(7). On the other hand, the conversion of sulfur in run-of-mine
coal to hydrogen sulfide does not appear to be a first-order process since the curves for this material in Figures 3 and 4 are
nonlinear. Because the run-of-mine coal contained large amounts
of both pyritic and organic sulfur, the nonlinear behavior could
have been caused by the superposition of reactions involving the
two sulfur species. Although the curves representing the formation rate of hydrogen sulfide were similar for both hydrogen and
nitrogen, it is apparent that for the same temperature and type
of coal, the rate was larger when hydrogen was used. This is only
natural since the rate should depend on the hydrogen concentration, and when pure nitrogen was fed, any hydrogen had to come
from the decomposition of the coal itself.

When an oxygen-bearing gas was used to treat coal, sulfur
dioxide was usually the major noncondensible sulfur compound in
the off-gas, but significant amounts of hydrogen sulfide were
also present. The formation rate of sulfur dioxide during several
runs made with an oxidizing gas is shown in Figure 5. For each
run two distinct peaks in the sulfur dioxide formation rate were
observed. The first peak might have been caused by devolatilization and oxidation of volatile sulfur compounds including hydrogen
sulfide. After the initial degassing of coal had subsided, oxygen
could penetrate the coal more readily and react with embedded
pyrites leading to the second peak. Then as the oxidation rate of
pyrites became limited by the diffusion of oxygen through an increasing layer of reaction products such as iron oxide, the rate
subsided. The difference in the behavior of the two types of coal
further supports this theory. Thus for deashed coal with a relatively small pyrite content, the second peak was much smaller
than for run-of-mine coal.

Analysis and Conclusions

The results of this study confirmed that it is possible to
remove substantial amounts of sulfur from pulverized bituminous
coal in a fluidized bed reactor operated at elevated temperatures.
However, for the type of coal used in this study, the removal of
sulfur is accompanied by a substantial loss of volatile matter.
Both the degree of desulfurization and extent of devolatilization
are strongly influenced by temperature. The composition of the
fluidizing gas appears to have more effect on the removal of
pyritic sulfur than on the removal of organic sulfur and volatile
matter in the 240°-400°C range. Thus an oxygen-bearing gas
appears more effective for removing pyritic sulfur than a
hydrogen-bearing gas, and nitrogen is completely ineffective. On
the other hand, the removal of organic sulfur appears due mainly
to pyrolysis and devolatilization and is not a strong function
of the treatment gas composition. Since a significant part of
the coal is volatilized, the recovery and utilization of the

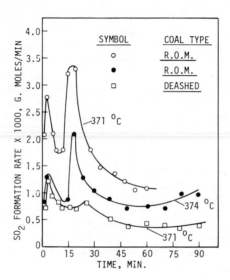

*Figure 5. Rate of SO₂ formation during
treatment with gas containing 10% oxygen*

volatile products is important.

Although a number of industrial process alternatives based on the fluidized-bed method of desulfurization are conceivable, only two will be considered here. One alternative involves treating pulverized coal in a continuous flow system with air or air diluted with recycled off-gas to remove pyritic sulfur and organic sulfur. This approach is indicated for coals containing finely disseminated pyrites which can not be removed by physical separation. It is conceivable that sufficient heat would be generated through oxidation to sustain the process. However, the off-gas would be diluted with nitrogen and would have a low heating value. Also the sulfur dioxide present in low concentration would be difficult to extract. On the other hand, the light oil in the off-gas would be relatively easy to remove, and there would be no tar to contend with. A second alternative involves treating coal in a flow system with recycled off-gas which has been desulfurized and heated. This approach is indicated for coals with important amounts of organic sulfur but little pyritic sulfur. The off-gas would be rich in hydrogen and methane and would have a relatively high heating value. Hydrogen sulfide in the gas would be relatively easy to remove, but the tar also present would create more of a problem than the light oil produced under oxidizing conditions. In the case of either alternative, the clean fuel gas would be used together with a char product.

While the methods applied in this study did not reduce the sulfur content of the selected coal to the point where the product would meet present air pollution control standards, further improvement in methodology is possible. From the published results of other workers ([1,2]), it is likely that either reducing the particle size or increasing the temperature would be beneficial, although increasing the temperature would remove more volatile matter as well as more sulfur. Also coals which initially contain less sulfur or are of a higher rank than the one selected could possibly benefit more from this type of treatment.

Achnowledgement

This work was sponsored by the Iowa Coal Project and conducted in the Energy and Mineral Resources Research Institute at Iowa State University.

Literature Cited

1. Jacobs, J. K., Mirkus, J. D., Ind. Eng. Chem. (1958), 50:24-26.
2. Sinha, R. K., Walker, P. L., Fuel (1972), 51:125-129.
3. Block, S. S., Sharp, J. B., Darlage, L. J., Fuel (1975), 54:113-120.
4. Cernic-Simic, S., Fuel (1962), 41:141-151.
5. Maa, P. S., Lewis, C. R., Hamrin, C. E., Fuel (1975),

         54:62-69.
6.   Snow, R. D., Ind. Eng. Chem. (1932), 24:903-909.
7.   Yergey, A. L., Lampe, F. W., Vestal, M. L., Day, A. G.,
      Fergusson, G. J., Johnson, W. H., Snyderman, J. S.,
      Essenhigh, R. H., Hudson, J. E., Ind. Eng. Chem., Proc. Des.
      Dev. (1974), 13:233-240.
8.   Batchelor, J. D., Gorin, E., Zielke, C. W., Ind. Eng. Chem.
      (1960), 52:161-168.
9.   Gray, C. A., Sacks, M. E., Eddinger, R. T., Ind. Eng. Chem.
      Prod. Res. Dev. (1970), 9:357-361.
10.   Zielke, C. W., Curran, G. P., Gorin, E., Goring, G. E.,
      Ind. Eng. Chem. (1954), 46:53-56.

INDEX

INDEX

A

Absorption isotherms 234
Acetone 157
Acid
  /leach/hydrodesulfurization 280
  -leached char, inhibition isotherms
    for 288
  -leached flash pyrolysis char, hydro-
    desulfuriaztion of 287
Air 305
  flow rate on flotation, effect of 77
  regeneration by 156
  table 46
  –water oxydesulfurization process .. 171
Alkali
  concentration on sulfur extraction
    from pyrite, effect of 188
  in HTT coals, sulfur capture by 202
  impregnation with 201
  leaching ISU coal with 190
Alkaline solutions, desulfurizing
  coal with 182
American Electric Power Service
  Corp. 153
Ammonia
  anhydrous 83
  breakage 58
  concentration, sulfur removal as a
    function of 176
  /oxygen system, sulfur removal
    from coals by 173
Ammonium carbonate 186
Appalachian region coals 37, 143
Arsenic 151
Ash 58
  content 308
  of ISU coal after leaching 193
  overall yield vs. 94
  vs. recovery curves 59
  reduction 92
Ashing for the direct determination of
  organic sufur, plasma 20
Ashing technique, low-temperature 19, 25
ASTM methods, standard 16
Atomic adsorption spectrometry,
  pyritic iron by 16
Autoclave 184
  high-pressure stirred 174

B

Batac jig 44
Batch
  experiments 164
  reactor runs with specified feedstock 277
  reactor tests 273
  sulfur removal in 278
Beneficiation
  on coal reserves 38
  by a float/sink technique 308
  kaolin 122, 129
  methods, coal 83
Benzene 166
BET method 222
Black water studies 53
Bomb apparatus, small 61
Btu loss from coals 180
Btu recovery 189
Bulk density 104
Bureau of Mines coal preparation
  program 48

C

Cadmium 151
Calcium 12
  carbonate 189
  hydroxide 189
  sulfate 157, 170
Capistrano test site, TRW's 153
Carbon 12
  from a char, initial loss of 227, 229
  dioxide 313
  impurity-free 236
  loss from coals 180
  monoxide 313
  –sulfur bond cleavage 208
Carbonization 305
  /desulfurization of Illinois coal,
    fluid-bed 248
  first-stage 254
  material flow and weight distribu-
    tion during staged 264
  PDU 250
    for clean coke process 251
    product char from 261
  second-stage 258
Carbonizer, fluid-bed 251, 265
Carbonyl sulfide 313

Carousel separator 122, 123
Cash flow financing method (DCF),
    discount 126
Cellular components 3
Cellular features of coalified plant
    material 5
Channel sample 14
Char 311
    from carbonization PDU, product .. 261
    composition 282
    desulfurization 221, 225, 235, 280
    electron probe analysis for 225
    filter-paper 238, 241
    flash pyrolysis 281
    forms of sulfur in derived 238
    hydrodesulfurization 280
        of flash pyrolysis 284, 285, 287
    Illinois 225
    inhibition isotherms for flash
        pyrolysis and acid-leached 288
    initial loss of sulfur and carbon
        from 227, 229
    oxygen concentration in synthetic .. 242
    pore surface area of 237
    sulfidation of previously
        desulfurized 235
    sulfidation of synthetic 233
    sulfur
        chemisorption on 244
        content, effect of temperature
            and residence time on 264
        effect of H$_2$S in fluidizing gas on 260
    after treatment, surface area of 231
Chemical 84
    analysis methods 85
    cleaning, combined physical and .. 192
    comminution 58, 59, 61, 84
        on overall results, effect of 97
    desulfurization of coal 52, 153, 173
    kinetics and mechanism of pyrite 182
    leaching 192
    treatment, combined physical and .. 195
    treatments, innovative 52
Chemicals, effective 63
Chemisorbed sulfur 240
Chemisorption on chars, sulfur 244
Chemisorption isotherms 245
Chlorination 207
    kinetic data for 213
    sulfur and chlorine in coal during .. 214
Chlorine in coal during chlorination .. 214
Chlorinolysis, coal desulfurization by
    low-temperature 206
Chlorinolysis data for bituminous
    coal desulfurization 212, 215
Clay, the wet beneficiation of kaolin .. 122
Clean coke process 248, 249
    carbonization PDU for 251

Cleaning
    bituminous coal and lignite,
        mechanical 36
    chemical 192, 198
    circuit, novel fine-size 46
    combined physical and chemical 192
    economic evaluation of physical
        coal 51
    float–sink 153
    Gravichem 153
    mechanical 36
    multistream coal 42
    physical 145, 192
    precombustion 121
Closed water circuits 46
Coal(s)
    with alkali, leaching ISU 190
    analysis data of tested 210
    analysis, starting 175
    Appalachian 39, 143
    availability, effect of crushing and
        cleaning on sulfur 41
    beneficiation methods 83
    bituminous 4, 36, 109, 110, 212, 215, 305
    Btu and carbon loss from 180
    carbonization of 248
    chlorination, kinetic data for 213
    during chlorination, sulfur and
        chlorine in 214
    cleaning, economic evaluation of
        physical 51
    composition 282
    contact vessel, coarse 162
    dechlorination data for bituminous 212
    desulfurization 221, 280, 290
        with alkaline solutions containing
            dissolved oxygen 182
        by ammonia/oxygen system 173
        chemical 52, 153, 173
        chlorinolysis data for
            bituminous 212, 215
        flotation process for 70
        in a fluidized-bed reactor 248, 305
        by HGMS 128, 130, 132, 133
        by high-intensity high-gradient
            magnetic separation 121
        of Illinois 223, 248
        kinetic data for 213
        laboratory 214
        by low-temperature chlorinolysis 206
        magnetic 112, 125, 129
        Meyers process for 143
        oxidative 164
        rate constants for 231
        test plant status 153
    devolatilization, resident
        time on 227, 257, 258
    dewatering 46, 52, 53

Coal (*continued*)
the direct determination of organic
sulfur in raw 22
dry table - pyrite removal from 101
drying unit operation 159
electron probe analysis for 225
flotation 72–74
framboid, pyrite in Iowa 9
gaseous treatment of 290
gasification at 223
HTT (*see* HTT coal)
hydrodesulfurization 267, 280
from ICO mines 86
Illinois No. 6 ....59, 61, 112, 174,
223, 248, 254, 275
Iowa 185, 291
ISU (*see* ISU coal)
Jude mine 86, 297, 309
kinetic data for chlorination
and desulfurization 213
after leaching, ash and sodium
content of ISU 193
leaching experiments 187
after leaching, sulfur content of ISU 191
liberating the mineral matter from .. 58
low-sulfur 40
Lower Kittanning 158
magnetic desulfurization of 112, 129
Martinka mine 148, 153
mechanical cleaning and thermal
drying of bituminous 36
Meyers process for desulfurization
of U.S. 143
microstructure 3
morphology of untreated
and HTT 200
nitration of the 166
organic sulfur removal 168, 297
oxidation oxygen consumption for .. 178
oxidation technique applied to 30
oxidative desulfurization of 164
oxidizer, continous 253, 255
oxydesulfurization 169
oxygen uptake by 178
physically cleaned 145
Pittsburgh seam 228
Pennsylvania mine 274
West Virginia mine 272
precombustion cleaning of 121
preparation 35, 70
process development facility,
central 55
prep/FGD combination study 51
pretreatment, sulfur removal
indicating effect of 269
process, hydrothermal 198, 200
process, morphology of untreated
and HTT 200

Coal (*continued*)
processing, coarse 159
product and recycle, bituminous 110
product and recycle, subbituminous 109
pyrite
with ferric sulfate, oxidation of 156
flotation, two-stage 50
removal by 167
by a reaction of leaching, extraction
of sulfur from 141
refluxed with cyclohexane 20
refuse, fine 53
reserves, beneficiation on 38
resources 38, 39
run-of mine 318
samples, composition of 292
samples in thermal analyses of 204
size consist of Illinois No. 6 59
studies 27
subbituminous 107, 109
sulfidation of 221, 233
sulfur
capture by alkali in HTT 202
direct determination of organic
forms in 16, 37, 145, 238
removal of organic 168, 297
removal by reaction and leaching 141
thermogravimetic analyses of
raw and HTT 204
toxic metals extracted by hydrothermal treatment of 203
trace elements in 54
trace metal removal from 215
type, effect of 63
washability analysis of 88
washing 144
/water slurry by HGMS,
desulfurization of 125, 130, 132
Western Kentucky No. 9 271
yield 88
vs. liberated pyritic sulfur
recovery 80
Coalified plant fibers 4
Coalified plant material, cellular
features of 5
Coarse coal contact vessel 162
Coarse coal processing 159
Coke process, clean 248, 249, 251
Collection procedure 24
Combustion behavior 201
Combustion, direct-fired 207
Comminuted coal, nitrogen content
of chemically 67
Comminution, chemical 58, 59, 61, 84
Comminution, mechanical 83
Conditions, operating 135
Control circuit, dense-medium
specific gravity 45

Control circuit, novel 44
Cost(s)
  of desulfurization of coal/water
    water slurry by HGMS,
    estimated 130
  of desulfurization of coal/water
    slurry by HGMS sensitivity
    analysis unit 128
  estimation 121
  of HEMF operating 118
  investment 136
  of magnetic desulfurization of
    coal/water slurry 125
  of magnetic desulfurization of
    liquefied coal 129
  operating 136
  processing 156
  of pyrite removal processes,
    capital and unit 130
  of solid–liquid separation methods,
    capital and unit 134
Cross-sectional areas 243
Crushing 84
  mechanical 59, 61, 83
  roll 84
  stage 39
  on sulfur coal availability effect of .. 41
Crystals 13
Crystallite sizes 236, 246
Crystallinity of filter-paper chars 238
Cyclohexane 20, 166
Cyclone, dense-medium 46

D

DBT sulfone 166
Decalin 166
Dechlorination 209
  data for bituminous coal,
    preliminary 212
Decomposition products, volatile 310
Dense-medium specific gravity
  control circuit 45
Density, bulk 104
Desulfurization
  coal (see Coal desulfurization)
  char (see Char desulfurization)
  flue gas 121
Devolatization, coal 227, 257, 258
Dewatering coal 46, 52, 53
Dibenzothiophene 165
Discharge distribution, dry table 108
Discount cash flow (DCF)
  financing method 126
Disposal, landfill-type 53
Dry table
  discharge distribution 108
  principle 102
  pyrite removal from coal 101

Drying of bituminous coal and
  lignite, thermal 36

E

Economic evaluation of physical
  coal cleaning 51
Economics 118
Electrode graphite 236
Electromagnet 121
Electron microprobe x-ray analyzer .. 12
Electron probe analysis for coal and
  char desulfurization and gasifi-
  cation 225
Electrostatic precipitators 125
Elemental sulfur 156
  recovery 159
Emissons
  controlling sulful oxide 37
  of low-sulfur coals, sulfur 201
  trace metal 201
Energy dispersive x-ray analysis 3
Enviro-clear thickener 49
Equipment performance evaluations .. 54
Errors 17
Ethane 313
Evacuation and coal rank, effect of 66

F

Feedstock, batch reactor runs with
  specified 277
Ferric
  oxide 183
  sulfate 183
    leaching 143, 153
    oxidation of coal pyrite with 156
    regeneration of 159
Ferrous sulfate 156, 183
Ferrous sulfide 306
Financing method, the discount
  cash flow (DCF) 126
Fine-size cleaning circuit, novel 46
Filtration of leach solution 159
Filtration, precoat 135
First-order process 317
First-order reaction 288
Five-stage pressure vessel 161
Flammability 203
Flash pyrolysis char 281
  hydrodesulfurization of 284, 285, 287
  inhibition isotherms for 288
Float/sink
  cleaning 153
  separation 143
  technique, benefication by a 308
Flotation
  froth 50, 84, 98
  kinetics, coal 73

Flotation (*continued*)
of Lower Kittanning and Pittsburgh
seam coals 75
operating conditions, effect of 77
process for desulfurization of coal .. 70
of pyritic sulfur, effect of fuel oil on 76
rate of Pittsburgh seam coal 74
systems, pyrite in coal 72
two-stage coal–pyrite 50
Flue gas desulfurization 121
Fluid-bed carbonizer 251
continuous pressurized 265
Fluid-bed carbonization/desulfuriza-
tion of Illinois coal 248
Fluidized-bed 273
reactor, desulfurization of
coal in a 248, 305
Fluidizing gas on char sulfur, effect
of H₂S in 260
Framboids 6
Froth flotation 50, 84, 98
Frother concentration, effect of 77
Frother systems, effect of 75
Fuel oil 83
on flotation of pyritic sulfur,
effect of 76
Fuels, solid 199

G

Gas
on char sulfur, effect of H₂S in
fluidizing 260
containing hydrogen, H₂S formation
during treatment with 312
/solid reactions, sulfur removal by .. 219
Gaseous treatment of coal 290
Gases, oxygen-bearing 310
Gasification of dried Illinois
No. 6 coal 223, 225
Gasification, secondary 232
Graphite, electrode 236
Graphite, pyrolitic 236
Gravichem cleaning 156
Grab and run process 79
Gravity separation 84, 97, 153, 192
Grinding, mechanical 83, 84
Grinding on overall results, effects of 96
Grinds of ISU coal 194
Gypsum 8
crystals 11

H

H₂S 260, 263, 312
H₂SO₄ 167, 170
Hydrogen 270, 293, 305
formation rates of 314
H₂S formation during treatment
with gas containing 312

Hydrogen (*continued*)
peroxide 313
sulfide 306
sulfur released in 294
Hydrogenation 125
Hydrolysis 209
Hydroperoxide 166
Hydrothermal coal process 198
morphology of 200
toxic metals extracted by 203
Heating value 189, 314
Heats of chemisorption 243
HEMF, operating costs of 118
HEMF unit 116
HGMS (high-gradient magnetic
separation) 51, 121, 122
estimated costs of desulfurization of
coal/water slurry by 130
sensitivity analysis unit costs of
desulfurization of coal/water
slurry by 128
High-intensity high-gradient magnetic
separation, desulfurization of
coals by 121
High-sulfur bituminous coals 84
Homer City preparation plant 42
HTT coal
morphology of 200
sulfur capture by alkali in 202
thermogravimetic analyses of 204
Hydroclone, separator 131
Hydrodesulfurization
acid leach/ 280
of acid-leached flash pyrolysis char 287
of coal char, improved 280
of coals 267, 290, 291
of flash pyrolysis char 284, 285
removal of sulfur by 219

I

ICO mines, coal from 86
Ignition temperature 203
Illinois char, desulfurization and
gasification of 225
Illinois coal 59, 61, 112, 174,
223, 248, 254, 275
Impeller speed on the flotation,
effect of 77
Impurities, level of 95
Inhibition isotherms for flash pyrolysis
and acid-leached char 288
Ion chromatograph 24
Iowa coals 185, 291
Iron 12, 156
by atomic adsorption spectrometry,
pyritic 16
oxides 182

Iron (*continued*)
pyrites 83, 185
pyrrhotite equilibrium 234
reduced 232
sulfates 143, 156
   crystallization 159
   washing of residual 159
Isotherm
absorption 234
chemisorption 245
Langmuir 240
ISU coal 191
with alkali, leaching 190
grinds of 194
after leaching, ash and sodium
   content 193
after leaching, sulfur content of 191

J

Jude strip mines, coal from 86, 297, 309

K

Kaolin
beneficiation 122, 129
lattice 17
Kinetic(s)
data for chlorination and
   desulfurization 213
equation 148
method, nonisothermal 306
process chemistry 156

L

Laboratory coal desulfurization 214
Laboratory processing 210
Lamella thickener 49
Landfill-type disposal 53
Langmuir isotherm 240
Leach
rate 157
solution, ferric sulfate 153
solution, filtration of 159
Leaching
chemical 192
experiments, coal 187
extraction of sulfur from coal by 141
ferric sulfate 143
ISU coal with alkali 190
of pyritic sulfur, pressure 159
–regeneration, simultaneous 159
sulfur content of ISU coal after 191
Lead 151
Lignite
deposits 53

Lignite (*continued*)
mechanical cleaning and thermal
   drying of 36
sodium content of 53
Limestone 170
Lithium aluminum hydride 18
Low
-sulfur char 248
-sulfur coals, sulfur emissions of 201
-temperature ashing technique 19
Lower Kittanning coal 75, 158

M

Maceral grouping 3
Magnetic
desulfurization
   of coal/water slurry 125
   of Illinois Basin coals 112
   of liquefied coal 129
separation (HGMS) 51, 121
Magnex process 129
Mags 122
Manganese 151
Marcasite 6
Martinka mine coal 148, 153
Mass balance, material flows 263
Mass spectrometer 306
Material balances, sulfur 299
Material flows 263
during staged carbonization, ... 264
Metal extraction 201
Metals extracted by hydrothermal
   treatment of coal, toxic 203
Methane content on the rate of sulfur
   removal, effects of 221
Meyers process for desulfurization
   of U.S. coal 143
Microchemical features 6
Microchemical studies 4
Milling, ball 84
Mineral
-free basis 18
inclusions 6
liberation 63
matter from coal, chemical
   comminution to liberate 58
residue separation method 125

N

Nickel 151
Nitration of coal 166
Nitrogen 293, 305
content of chemically comminuted
   coal 67
dioxide 166
H_2S formation during pyrolysis in 312

Nitrogen (*continued*)
mixture, oxygen– 296
sulfur released in 294
Nonisothermal kinetic method, ... 306

O

Oil
agglomeration 85
concentration, pine 78
on flotation of pyritic sulfur, effect
of fuel 76
fuel 83
light 310
Optical microscopy 222
Organic solvents 157
Organic sulfur 165, 173, 182, 207, 222,
262, 270, 283, 292, 305
from oxidation, repeatability runs of
the analyses of 28
plasma ashing for the direct
determination of 20
in raw coals, the direct
determination of 22
removal from coals by
oxydesulfurization 165, 168
removal vs. sample weight loss 297
Oxidants 166
Oxidation
of coal pyrite with ferric sulfate 156
organic sulfur from 28
product 183
of pyritic sulfur, selective 305
study of FeS_2 27
technique, low temperature 23
technique to various coals, application of the 30
Oxidative desulfurization of coal 164
Oxidized Illinois coal, analysis of 254
Oxidizer, continuous coal 253, 255
Oxydesulfurization
of coals 169
organic sulfur removal from
coals by 168
process, air–water 171
pyrite removal from coals by 167
Oxygen
-bearing gases 310
concentration in synthetic chars 242
consumption 175
for coal oxidation 178
minimum 177
content of filter-paper chars after
sulfidation 241
content of chars, initial 238
desulfurizing coal with alkaline
solutions containing dissolved .. 182
–nitrogen mixture 296

Oxygen (*continued*)
regeneration by 156
on the sulfur absorption,
influence of 239
system, sulfur removal from coals
by ammonia– 173
uptake by coal 178

P

Particle size on sulfur extraction from
pyrite, effect of 188
Particles, volume of 193
PDU
carbonization 250, 251
product char form 261
scale tests 276
studies 248
Phases, inorganic 3
Physical and chemical cleaning,
combined 192, 195
Physical methods for removing
sulfur from coal 33, 145
Pine oil concentration 78
Pittsburgh seam coal 74, 75, 228
Plant
debris 7
fibers, coalified 4
material, cellular features of
coalified 5
Plasma ashing for the direct determi-
nation of organic sulfur 20
Pore surface area of char 237
Precipitators, electrostatic 125
Precision 28
Precoat filtration 135
Precombustion cleaning of coal 121
Pre-evacuation, effect 65
Preparation plant, Homer City 42
Preparation process development
facility, central coal 55
Pressure
leaching of pyritic sulfur 159
on the rate of sulfur removal,
effects of 65, 221
vessel, five-stage 161
Pretreatment 268
Price for desulfurization of SRC feed
coal by HGMS, steam 133
Process chemistry, kinetics and
scheme of 156
Process, clean coke 248, 251
Processing costs 153
Pulverizing 84
Pyrite(s) (FeS_2O) 222, 270, 293
in coal flotation systems 72
distribution 3, 12, 13
effect of alkali concentration on
sulfur extraction from 188

Pyrite (*continued*)
  effect of particle size on sulfur
    extraction from 188
  effect of reaction time on sulfur
    extraction from 191
  extraction rates 157
  finely dispersed 17
  flotation, two-stage coal 50
  in Iowa coal 10, 11
  iron 83, 185
  leaching experiments 185
  proportion of 13
  reactivities, of the 148
  removal 71
    from coal, dry table 101
    from coals by oxydesulfurization 167
    processes, capital and unit costs of 130
    system, pyrrhotite/ 236
Pyritic sulfur 39, 58, 159, 165,
                              173, 182, 208, 222
  overall yield vs. 93
  reduction 91
  removal 129, 146, 147
    rate constants for 148, 149
  selective oxidation of 305
Pyrolitic graphite 236
Pyrolysis 305
  H_2S formation during 312
  removal of sulfur by 219
Pyrrhotite (FES) 18, 222, 232
  equilibrium, iron/ 234
  highly magnetic 125
  /pyrite system 236

R

Rate constant 228
  for desulfurization 231
  for pyritic sulfur removal 148, 149
Raw coals, thermogravimetic
  analyses of 204
Reactivity, increased 203
Reaction
  conditions, effect of 63
  extraction of sulfur from coal by 141
  first-order 228
  procedure 24
Recovery curves, ash vs. 59
Recovery curves, sulfur vs. 61
Recycle, bituminous coal product and 110
Recycle, subbituminous coal
  product and 109
Reduction potential 39
Reduction, sulfur 144
Refuse, fine coal 53
Refuse pond stabilization study 53
Regeneration 156
  of ferric sulfate 159

Regeneration (*continued*)
  by oxygen or air 156
  simultaneous leaching– 159
Repeatability runs of the analyses of
  of organic sulfur 28
Resiliency 104
Roll crushing 84

S

Sample, representative and
  homogeneous 18
Sandwich screen concept 46
Scanning electron microscope 3
Semicontinuous experiments 165
Sensitivity analysis unit costs,
  desulfurization of coal/water
  slurry by HGMS 128
Separation
  float-sink 143
  gravity 84, 97, 153, 192
  magnetic (HGMS) 121
  method 135
    capital and unit costs, solid–liquid 134
    mineral residue 125
Separator
  Carousel 122
  hydroclone 131
  matrix loading characteristics 127
Shape 104
Silica sand 308
Size 104
  consist of Illinois No. 6 coal 59
  effect of starting 65
  reduction methods 84
Sizing 46
Sodium
  carbonate 166, 186
  content of ISU coal after leaching .. 193
  content of lignite 53
  hydroxide 166, 183, 186
  phosphate 186
Solid
  fuels 199
  –liquid separation methods,
    costs of 134, 135
  removal of sulfur by gas/ 219
Solvents, organic 157
Species, ionic 17
Specific gravity control circuit,
  dense–medium 45
Stabilization study, refuse pond 53
Stage crushing 39
Steam 305
Sulfate 156
  analyses 24
  conversion to 208

Sulfate (*continued*)
ferric and ferrous 167
sulfur 6, 156
Sulfidation
of coal 221
char and synthetic chars 221, 233
oxygen and sulfur contents of
filter-paper chars after 241
of previously desulfurized char 235
Sulfide 270
fusinite, polycrystalline 10
inorganic 311
loss of 18
occurrence of 17
sulfur 283
Sulfone, DBT 165, 166
Sulfur (*see also* Pyritic sulfur, Organic
sulfur)
absorption of 234
absorption, influence of oxygen
on the 239
capture by alkali in HTT coals 202
change in forms of 262
for a char, initial loss of 227
chemisorbed 240
chemisorption on chars 244
in coal 37
during chlorination 214
coal availability, effect of crushing
and cleaning on 41
from coal by a reaction and leach-
ing, extraction of 141
content 115
of chars, initial 238
effect of temperature and resi-
dence time on char 264
effect of temperature and residence time on semichar 257
of filter-paper chars after
sulfidation, ... 241
of ISU coal after leaching 191
determination of forms 1, 16
dioxide 313, 314
distribution 311
in dried coal, forms of 238
effect of H_2S in fluidizing
gas on char 260
elemental 156, 183
direct determination of 20
emissions of low-sulfur coals 201
extraction 199
from pyrite, effect of alkali concentration and particle
size on 188
from pyrite, effect of reaction
time on 191
forms of U.S. coals 145
to H_2S, conversion of forms of 263

Sulfur (*continued*)
inorganic 262, 292
material balances 299
organic (*see* Organic sulfur)
oxide emissions, controlling 37
pyritic (*see* Pyritic sulfur)
reactions, parameters for 300, 301
vs. recovery curves 61
recovery, elemental 159
reduction 91, 115, 144
potential 38, 39
released in hydrogen 294
released in nitrogen 294
removal
in batch reactor tests 278
from coals by ammonia/
oxygen system 173
effect of temperature on 295
effects of temperature, pressure,
and methane content on the
rate of 221
effect of holding time on 295
as a function of ammonia concentration and time 176
indicating effect of coal pretreatment 269
pyritic (*see* Pyritic sulfur removal)
by pyrolysis, hydrodesulfurization, and other gas/solid
reactions 219
removed vs. temperature, percent .. 278
scavenger 201
sulfate 6, 156
sulfide 283
–sulfur bond 208
transformation of 299
Sulfuric acid 156, 167, 183
Surface
area 240
of char after treatment 231
of char, pore 237
phenomena in coal dewatering 53
roughness 104
Surfactant solution as dewatering
aid, hot 52

T

Tails 122
Tar, black 310
Technical feasibility 183
Temperature
on char sulfur content, effect of 264
on coal devolatilization, effect of 257
ignition 203
percent sulfur removed vs. 278
on the rate on sulfur removal,
effects of 221

Temperature (*continued*)
on semichar sulfur content, effect of   257
on sulfur removal, effect of   293
Test
evaluation   79
plant design and operation   159
plant status, coal desulfurization   153
Tetralin   166
Thermal analyses of coal samples in   204
Thermal efficiency   179
Thermobalance tests   270
Thermogravimatic analyses of raw
and HTT coals   204
Thickener, Enviro-clear   49
Thickener, Lamella   49
Time
on char sulfur content, effect of   264
on coal devolatilization, effect of
residence   257
on semichar sulfur content, effect
of residence   257
on sulfur extraction from pyrite,
effect of reaction   191
on sulfur removal, effect of holding   295
sulfur removal as a function of   176
Trace element
in coal   54
content   143
removal   150, 151
Trace metal emissions   201
Trace metal removal from coal   215
Transmission electron microscope   3
Treatment, combined physical and
chemical   195
TWR's Capistrano test site   153
Toxic metals extracted by hydrothermal treatment of coal   203

V

Volatile matter   309

W

Washability
analysis of coal   88
characteristics   38
comparisons   58
Washing
coal   144
with dilute acid   192
of residual iron sulfate   159
with water   192
Water
content, effect of   65
oxydesulfurization process, air–   171
slurry, coal–   125
studies, black   53
Wave-length dispersive analysis   12
Weight
distribution during staged
carbonization   264
loss for deashed Jude coal sample   297
yield of product   90

X

X-ray analysis, energy dispersive   3
X-ray analyzer, electron microprobe   12

Y

Yield vs. ash content, overall   94
Yield vs. pyritic sulfur, overall   93

Z

Zinc   151